T0254681

SpringerBriefs in Applied Sciences and Technology

Safety Management

The SpringerBriefs in Safety Management present cutting-edge research results on the management of technological risks and decision-making in high-stakes settings.

Decision-making in high-hazard environments is often affected by uncertainty and ambiguity; it is characterized by trade-offs between multiple, competing objectives. Managers and regulators need conceptual tools to help them develop risk management strategies, establish appropriate compromises and justify their decisions in such ambiguous settings. This series weaves together insights from multiple scientific disciplines that shed light on these problems, including organization studies, psychology, sociology, economics, law and engineering. It explores novel topics related to safety management, anticipating operational challenges in high-hazard industries and the societal concerns associated with these activities.

These publications are by and for academics and practitioners (industry, regulators) in safety management and risk research. Relevant industry sectors include nuclear, offshore oil and gas, chemicals processing, aviation, railways, construction and healthcare. Some emphasis is placed on explaining concepts to a non-specialized audience, and the shorter format ensures a concentrated approach to the topics treated.

The SpringerBriefs in Safety Management series is coordinated by the Foundation for an Industrial Safety Culture (FonCSI), a public-interest research foundation based in Toulouse, France. The FonCSI funds research on industrial safety and the management of technological risks, identifies and highlights new ideas and innovative practices, and disseminates research results to all interested parties.

For more information: https://www.foncsi.org/.

More information about this series at http://www.springer.com/series/15119

Claude Gilbert · Benoît Journé
Hervé Laroche · Corinne Bieder
Editors

Safety Cultures, Safety Models

Taking Stock and Moving Forward

Editors
Claude Gilbert
Laboratoire PACTE, Science-Po Grenoble
CNRS/FonCSI
Grenoble, France

Benoît Journé
Université de Nantes
Nantes, France

Hervé Laroche
ESCP Europe
Paris, France

Corinne Bieder
Ecole Nationale de l'Aviation Civile
Toulouse, France

ISSN 2191-530X ISSN 2191-5318 (electronic)
SpringerBriefs in Applied Sciences and Technology
ISSN 2520-8004 ISSN 2520-8012 (electronic)
SpringerBriefs in Safety Management
ISBN 978-3-319-95128-7 ISBN 978-3-319-95129-4 (eBook)
https://doi.org/10.1007/978-3-319-95129-4

Library of Congress Control Number: 2018949053

This Springer imprint is published by the registered company Springer Nature Switzerland AG
The registered company address is: Gewerbestrasse 11, 6330 Cham, Switzerland

Foreword

The term 'safety culture' is encoded in the name of the Foundation for an Industrial Safety Culture (FonCSI) and, as such, reflects the fundamental purpose of our research foundation. But what actually lies behind these words? There are various definitions of 'safety culture' as a concept, and it is still the subject of theoretical debates and quarrels. With the view to improving both occupational and process or product safety, industry is presented with a wide offer of safety models, safety culture approaches, methods and tools.

What is the landscape more than 30 years after the term 'safety culture' appeared? Should safety culture be considered as a system of values or as a normative tool? Is it possible to change or improve safety culture? Should safety models be addressed from a prescriptive point of view or rather through an analytical perspective? Is there one 'best' safety model? What are the links between safety culture and models? How can we choose, given the tremendous offer in the safety culture ideas' market? What issues should industry address to go one step beyond regarding safety?

These are a brief summary of some of the questions FonCSI addressed in this second 'strategic analysis', an innovative research methodology which seeks to provide FonCSI's partners with high-level research results within a limited time. The final aim of the 'strategic analysis' is to produce both a state of the art of practices and operational axes of improvement for industries carrying out hazardous activities. Analysis occurred in a two-stage process. The first involved a small group of experts from different academic disciplines and various industrial sectors such as oil and gas, energy and transportation: the FonCSI 'strategic analysis' group. The group brainstormed, exchanged on a monthly basis over a one-year period on these questions and identified internationally recognized scholars on the topic of safety culture and models. The second stage included, in addition to the core group, the identified scholars who were invited to present their work and confront their viewpoints during a two-day residential seminar, the highlight of this project, held in June 2016.

This book reflects the rich debates that occurred not only at the conceptual level but also regarding the operational and political issues faced by high-risk industry when it comes to safety culture. After an introductive chapter detailing the expectations of FonCSI's industrial partners, the book presents the in-depth reflections conducted on the questions raised above and more. By displacing the usual reading grids, challenging the term safety culture and questioning the purpose and relevance of models, this book helps to dispel the 'safety cloud' (as it was called by the industrial members of the group) of concepts and approaches and proposes ways forward for at-risk industries. We encourage you to read it, share it and discuss it!

Caroline Kamaté
François Daniellou
FonCSI, Toulouse, France

Contents

Chapter 1
An Industrial View on Safety Culture and Safety Models

What to Choose and How in the Nebulous "Safety Cloud" of Concepts and Tools?

Olivier Guillaume, Nicolas Herchin, Christian Neveu and Philippe Noël

Abstract This chapter, co-written by the industrial members of the FonCSI "strategic analysis" group, gives an overview of the various contexts and histories of safety culture/safety models throughout the four industries represented, and summarizes the main questions and issues arising from an industrial point of view. In brief, in a context of high industrial risks—both in terms of process safety and occupational safety—two main topics emerge for discussion: (i) the question of the co-existence of several safety models: what to choose and according to what criteria from the panel of tools available? And (ii) the specific notion of "safety culture": what more does the concept bring, and how to apprehend it in complex industrial organisations? Eventually, the expression "safety cloud" is used to illustrate the overall feeling of confusion in the industrial world: the current perception is one of a nebulous offer of various models and tools, the choice of which appears difficult to rationalize and adapt to a company's specifics and local issues. As an introduction to more academic discussions, this chapter thus sets the tone and hopes to shed light on some unanswered industrial questions.

Keywords Industrial · High-risk · Safety · Safety cloud · Safety culture Safety models

O. Guillaume (✉)
EDF Lab, Paris-Saclay, France
e-mail: olivier.guillaume@edf.fr

N. Herchin
ENGIE Research & Technologies Division, Paris-Saint Denis, France

C. Neveu
SNCF Safety System Department, Paris-Saint Denis, France

P. Noël
TOTAL HOF Division, Paris-La Défense, France

C. Gilbert et al. (eds.), *Safety Cultures, Safety Models*,
SpringerBriefs in Safety Management,
https://doi.org/10.1007/978-3-319-95129-4_1

1 Introduction

In their contexts of high industrial risks, the four companies:

- EDF (Electricité de France), representing the nuclear sector
- SNCF (Société Nationale des Chemins de Fer Français) for the railway sector
- ENGIE (ex GDF SUEZ), a global energy company
- TOTAL, well-known major in the petrochemical sector

share the same concern with safety matters, striving to develop high safety standards that lead to mature safety cultures and the lowest accident rates.

This chapter aims to synthetize the main issues raised by their representatives within the FonCSI "strategic analysis" group in the field of safety culture and models. In Sect. 2, the context and specifics of each of the FonCSI member companies is presented, along with their main search regarding safety aspects. The third section seeks to produce a digest of these issues to provide a common core of questions and needs around the concepts of safety models and safety culture throughout industry.

2 Various Industrial Contexts Leading to Different Histories of Safety Models and Safety Culture Approaches

2.1 The Nuclear Industry: The Case of EDF

Safety culture is certainly one of the toughest topics in nuclear safety because it is a matter of improving human functioning in a very technical and regulated industry. Talking about safety culture requires us to keep in mind the accidents of Chernobyl, Tokai-mura or Fukushima, but also to recall the major events at Davis Besse, for example. In a Nuclear Power Plant (NPP), people have to take into account the diversity of situations, deal with multiple-choices and be prepared to face very rare situations. Finally, industrial safety covers a rule-based part and a managed safety aspect which can handle variability and the unexpected.

In EDF, several historical events allowed the safety culture to develop. After Three Mile Island, EDF implemented an independent safety line and regular safety assessments—in order to challenge the operational lines regarding safety. These devices developed questioning attitudes, cross visions, continuous improvement and made safety a priority.

The second step was after the Chernobyl accident where EDF brought human factor specialists into every NPP and engineering unit, in order to reinforce a technical, but also human & organisational approach to developing Human Factor knowledge and methods among managers and employees.

Thus, in the 90s INSAG[1] 4 (INSAG, 1991) & 13 (INSAG, 1999) were used as a foundation to develop the Safety Culture and six levers were developed in particular, in order to implement the Safety Culture.

In the years 2000, other lessons were learned from INSAG 4, including that Safety Culture is not only a matter of individual behaviour, but involves the entire management line. In order to support and to develop the crucial role of managers to improve safety, the nuclear division of EDF produced a safety management guideline, which described what is expected for each level of management and focused on key-principles: safety leadership, staff development and commitment, oversight and continuous improvement, and a crucial practice of “managers in the field”.

In 2013, the decision was taken to boost the EDF Safety Culture approach in light of the Fukushima accident, but also due to a huge renewal of generations and employees and the creation of new international guidelines. In order to boost collective thinking, the agency pyramid was used considering that Safety Culture is the product of the interaction between three dimensions—organisational, behavioural and psychological.

In 2014, a team including corporate and site staff, was set up tasked with building a common representation of Safety Culture for the nuclear divisions. Using the knowledge of international guidelines—the agency and WANO[2]—and, taking into account the EDF nuclear fleet features, this team described Safety Culture via six themes, divided into around thirty sub-themes and some one hundred items.

Then, the way to use the guidelines and to develop Safety Culture were organised into three pillars:

1. skills development, focusing on young recruits; with videos, tutorials to pass on history, and active techniques with case-studies and coaching;
2. daily communication on Safety Culture;
3. collective Safety Culture assessment in order to encourage debate, collective thinking and stepping back, because Safety Culture cannot be decreed.

Finally, the Safety Culture guideline is developed with dedicated site actions.

Safety Culture also depends on professions. The objectives are to discover what these sub-cultures are, with their related beliefs and assumptions, and to think about how they can fit with the dedicated guide and the other tools. Indeed, experience shows us that it is not always relevant to use a very detailed and formal safety culture because it does not fit with all professional safety cultures and can be hard to manage if there are too many items.

In order to overcome these difficulties, collective meetings are developed in EDF NPPs. In these meetings, a simple graphic representation of safety culture is used as a “projective object” to help people to imagine what the main characteristics of their own professional safety culture are. This representation also helps participants to

[1]International Nuclear Safety Group.

[2]World Association of Nuclear Operators.

perceive what the weaknesses of their safety culture are and what are the best ways to strengthen it. A simple formalized safety culture where its main items can be understood and accepted by everyone, becomes a "common language".

Moreover, these collective meetings can become spaces for debate where members of several professions (operators, maintenance technicians…) can explain their safety representations, their activities, their risks and methods for solving them.

The "projective objects" used in these meetings bring an interpretative flexibility to the concept of safety culture. It becomes a "boundary object" which can facilitate the understanding between professions and managers and can help them to coordinate their diversities in order to create consensual approaches.

2.2 *The Railway Industry: The Case of the SNCF*

2.2.1 Brief Presentation of the SNCF

In 2015, France's Rail Reform Act created the new SNCF Group, a unified public service company that now generates €31.4 billion of revenue in France and in 120 countries around the world. Today's SNCF consists of three state-owned industrial and commercial enterprises—SNCF, SNCF Réseau and SNCF Mobilités.[3]

About 155,000 employees work in the French railway sector of the SNCF Group. Many professions are concerned by safety issues: drivers, signallers, shunters, rolling stock maintenance staff, infrastructure maintenance staff, traffic dispatchers, conductors, etc.

2.2.2 Organisation and General Issues in Terms of Health and Safety

Two of the three enterprises of the SNCF Group are confronted with industrial risks:

- SNCF Réseau in its activities of maintenance and development of the national railway network as well as traffic dispatching;
- SNCF Mobilités in its activities of railway undertaking delivering transport services for passengers and freight loaders.

Thus, they hold a safety authorisation (Réseau) or a safety certificate (Mobilités) granted by the French national safety authority for railway, the EPSF.[4] The authorisation or certificate covers all the risks related to railway operations, vis-à-vis

[3]The industrial context taken into account in this chapter is composed of all the railway activities in France. It does not include other modes of transport operated by the SNCF Group nor the railway operations overseas.

[4]EPSF: Etablissement Public de Sécurité Ferroviaire.

passengers, workers, subcontractors, outsiders, freight or environment. Non-railway risks are regulated by the requirements of the labour code.

Inside the SNCF Group, safety functions are allocated according to this legal, industrial and organisational framework. SNCF Réseau and SNCF Mobilités are in charge of devising their own safety management system and implementing it after approval by EPSF. This responsibility implies a strong involvement of managers in safety matters. Managers are supported by safety specialists located in every major profession of the company (drivers, signallers, maintenance, etc.) and at every managerial level (frontline, middle and top management).

The safety division of SNCF, known as the "safety system division", provides the whole SNCF Group with common principles and a range of management tools. It also puts the expertise of its teams at SNCF Group's disposal for technical, organisational or managerial issues related to interfaces between the network and trains.

The serious accidents which occurred in recent years as well as a high rate of occupational accidents have led the SNCF Group to carry out an in-depth review of its safety policies and methods.[5]

There has been an examination of the top-down oriented managerial approach that is strongly focused on exclusive compliance with rules and the supposed positive impact of the sanction, as well as the place of safety in strategic and operational decision-making processes. The "in-silos" way of functioning has also been highlighted.

An apparent paradox emerged: although SNCF has been working for more than 20 years in the human and organisational fields, the main findings of the experts' committee diagnosis pointed out the lack of a "Human" dimension in the safety system.

2.2.3 Needs Going Forward

This fundamental reflexion led to an ambitious program to produce deep changes in the safety culture. The SNCF Group is engaged in a profound overhaul of its safety management that involves taking a different view on what builds safety on a daily basis, on the place of rules, on human performance, on the role of expertise. It is a question of transforming the safety culture of the 155,000 employees of the SNCF Group.

What are the levers and appropriate references in order to reach such an ambitious goal? Is a safety management the starting point or the awaited output? Or both? Does a good safety culture exist which would be obvious to everybody and could be used as a target for all employees?

[5]This review has been pushed by an international experts' committee which was appointed by the executive committee of the Group.

2.3 The Energy Industry: The Case of ENGIE

2.3.1 Brief Presentation of ENGIE

ENGIE stems from the fusion in 2008 of Gaz de France and SUEZ, grouping together natural gas activities and infrastructures on the one side, and energy services and power facilities on the other. Today, ENGIE is focused on its three core businesses of Electricity, Natural Gas and Energy Services to support and develop a new vision of energy for the world: sustainable energy available to everyone.

Acting as a major stakeholder in the international energy industry, the ENGIE Group employs more than 154,000 people worldwide with operations in 70 countries generating an annual revenue of €69.9 billion (2015).

2.3.2 Organisation and General Issues in Terms of Health and Safety

The ENGIE Group is organised into twenty-four Business Units (BU) managing their activities independently, and five transverse "Métiers" providing support on key activities. As part of the Corporate functions, the Health and Safety Directorate provides support and guidance for the BUs, with three core missions: (i) steering and promotion of Health and Safety culture, (ii) functional line support and facilitation, (iii) return of experience and oversight.

The Group's international presence leads to a multiplicity of contexts and cultures in the various countries of implantation. Its multiple activities result in wide disparities in terms of risks, and therefore risk management, be it process safety or health and safety matters. As a consequence, in order to embrace all the local contexts and activities' specifics, the Group's H&S policy is built around a common thread and toolbox with sufficient autonomy being given to the BUs to manage risks.

Lastly, two main company cultures still tend to co-exist after the recent fusion of Gaz de France and SUEZ, the latter showing for example more immediate application of the rules set at the corporate level than the former, being more prone to discussions and local reinterpretations. This historical trait is reinforced by the type of risks managed, major hazards arising from the gas infrastructures.

2.3.3 Culture and Safety Models: Several Approaches and Tools

Historically, many theories and concepts as well as tools in terms of health and safety have been used, mostly coming from external sources (academic work or consultants), that are today more or less attractive and have more or less proven their worth.[6]

[6]These include, for example, the Bird pyramid, Reason's Swiss cheese model, the root cause analysis… with the primary concern of preventing accidents.

In the early 2010s, an interest for human and organisational factors and the concept of "safety culture" started to emerge amongst ENGIE's safety functional line, leading to:

- human and organisational factors approaches, for example in GRDF (Gas Distribution Network BU) as early as 2008, with a consultant;
- safety culture approaches, between 2010 and 2013 with the help of ICSI[7] performing safety culture diagnoses in six entities (gas and energy services).

These approaches were backed by the Group's R&D center, which provided support on the post-diagnosis phase with various HOF tools such as HOF accident analysis methods, just sanction and reward policies, managerial visits, workshops on collective mindfulness, etc.

In parallel, the Group's H&S direction developed a toolbox of guidance and support on several axes such as life-saving rules and commitment, safety training, sub-contracting, etc., all of which contribute to the Group's main goal in terms of safety: "*make our safety and health culture evolve towards a proactive and shared culture.*"

2.3.4 Needs Going Forward

With the recent input of internal R&D and external consultants from the perspective of Human and Organisational Factors, the concept of "safety culture" has grown widely amongst the Group's safety functional line, mostly stemming from the ICSI diagnosis program. Indeed, the term has been taken over by the corporate level in its safety ambition: "*towards a proactive and shared safety culture*".

Yet, several questions remain for the Group in terms of approach. Should there be one single model for the whole Group leading to a certain level of standardization, or, given the complexity and variety of activities, should there be as many models as there are different activities?

The question also arises of the difference in treatment of process safety aspects and general health and safety aspects. Usually, these two aspects tend to be managed by different entities. Should it be different for the sake of enhanced performances? Or are we dealing with two separate dimensions?

2.4 *The Petrochemical Industry: The Case of TOTAL*

2.4.1 Energy Company

Total is a major energy player committed to supplying affordable energy to a growing population, addressing climate change and meeting new customer

[7]Institute for an Industrial Safety Culture (Toulouse, France).

expectations. With operations in more than 130 countries, Total is the world's 4th ranked oil and gas company and a global leader in solar energy with SunPower. With 96,000 employees committed to better energy, Total is an integrated oil and gas company, from upstream to downstream (exploration and production, refining, marketing and developing new energy for the future).

2.4.2 The Way to Reach a High Level of Safety Performance

After years of improvement in safety performance by dealing with technical aspects of safety and implementing Safety Management Systems, analysis of major accidents and high potential incidents has led to the conclusion that there were still aspects to improve in these two domains but also in the field of human behaviours.

More than 15 years ago, it was decided to establish a Safety Culture position at corporate level in order to start some programs for Human and Organisational factors integration. At the beginning, these programs were based on behavioural approaches but quickly moved to a Safety Culture approach: integration of key factors that influence the way people think and the way they act.

The Safety Culture program still consists of analysing the Safety Culture components of an affiliate in order to strengthen them and to improve its safety performance. This program, built with ICSI and a sociologist from North America, is based on a diagnosis tool, established with four types of Safety Cultures: fatalistic culture, trade culture, managerial culture and integrated culture. The integrated Safety Culture is considered as the target for Total entities in the world, which have to take into account the local contexts and specifics of their organisation and metier to establish their own program.

2.4.3 A Strong Safety Model Is Expected

Local communities/authorities and international stakeholders expect Total, as a major in the oil and gas industry, to establish a strong company safety model. Thus, the Total Safety standards and the Safety company model must be implemented to ensure a high level of risk management in every entity, whatever the affiliate, its location in the world, the branch and its metier. Such top-down programs, typical signs of a managerial Safety Culture model, are not always consistent with the variability and adaptation principles, which come from academics and have led to the Corporate Safety Culture program, including the fact that each affiliate has to establish its own Safety Culture plan adapted to local specificities and fulfilling local needs.

2.4.4 Culture Prospective Broader Than Safety

As soon as a Total affiliate starts to analyse and to strengthen its Safety Culture, it has to deal with leadership aspects of the management line, organisational aspects including sociological considerations like relationships between people and groups of people within the entity, management methods, involvement of employees, resources and competencies and also the specific context of the entity. These factors are key components influencing the Safety Culture but are components of the organisational culture of the Company or of the entity at local level.

Therefore, some managers in the company wonder if entities have to perform specific Safety Culture programs or if the company must establish a Company Management Model, including Safety and other areas of risk as well.

3 A Common Core of Questions and Needs Around the Concepts of Safety Models and Safety Culture Throughout the Industry

Considering these various industrial contexts, it is possible to gather a common core of questions and expectations around the topic of safety models and cultures. The following paragraphs try to provide a digest of the main issues at stake from an industrial perspective.

3.1 How to Make One's Way Through the Numerous (Safety) Models Available in the Academic or Consulting Worlds?

3.1.1 Co-existence of Several Safety Models: What to Choose and According to What Criteria, in the Nebulous "Safety Cloud" of the Academic and Consulting Worlds?

The first, dominant issue seems to be the multiple offer existing around safety models and tools. Various schools of thought exist within the academic world: to date, an exhaustive view—if ever possible to produce—seems to be lacking of the various models available. In addition, how does the concept of "safety culture" position itself in relation to other safety concepts or models, such as organisational models (for example Resilience, HRO, etc.)?

In short, it appears that companies are surrounded by some sort of nebulous "safety cloud" from which they pick up various tools and ideas, following the trends of the moment sometimes without much consistency. In this context, the question becomes: what to choose (in terms of content) from the different existing models/tools? and perhaps more precisely *how*, i.e. according to what criteria, should the relevant model(s) be chosen for one company?

Finally, the very purpose of safety models themselves are questioned: are (good) safety organisational models sufficient to ensure safety? In other words, can the target models used guarantee that safety results will be improved in the end? And if so, linking back to the specific concept of "safety culture", what is the relevance of using this notion? What more does it bring?

Ultimately, should the model used for diagnosis and analysis be differentiated from the target safety model, used as a goal? Indeed, using the same model as both a target and a diagnosis tool to monitor progress might lead to bias in the way safety is apprehended.

3.1.2 Should There Be a Global, Homogeneous, Model, or Several Models Adapted to Local Specific Features?

One of the most relevant question for the industry is whether there should be a single model of safety culture for one whole Group or if a firm should manage a variety of safety culture models. And if an industry has to deal with a diversity of models of safety culture, how should these be managed and how can they be articulated with a more formal one?

In-depth analysis of the Chernobyl accident by IAEA[8] experts (1986) showed a lack of safety principles in design and production pressure, non-respect of operating procedures or lack of preparation in crisis management. Meanwhile, the root cause was a lack of safety culture and personal dedication, meaning a lack of safety thinking and absence of an inherently questioning attitude. If the INSAG 3 of IAEA (1988) described Safety Culture as

> A very general matter, the personal dedication and accountability of all individuals engaged in any activity which has a bearing on the safety of nuclear power plants,

the Nuclear Industry had to wait for the INSAG 4 (1991) to detail the Safety Culture. This guideline considered Safety Culture as

> That assembly of characteristics and attitudes in organizations and individuals which establishes that, as an overriding priority, nuclear plant safety issues receive the attention warranted by their significance.

After this industrial definition, scholars tried to formulate their own version considering safety culture as the part of the organisational culture which influences attitudes and behaviors, increasing or decreasing risks (Guldenmund, 2000). Indeed, several scholars formalized safety culture as an interaction of three components: a psychological (and sociological) component gathering perceptions and values about safety; a behavioral component gathering workers and managers'

[8]International Atomic Energy Agency.

behaviors and attitudes; a structural component including formal and informal organisation.

Meanwhile, industrial managers decided that they could not remain at this level of generality and began to detail and formalize safety culture in their industry. Later, for the nuclear industrial sector, a number of institutions including WANO, IAEA, NRC[9] created a guidance document called "Common Language" (in 2014). This document listed and detailed the Safety Culture attributes sorted by categories and components. This "common language" and safety culture guidelines helped several nuclear operators to formalize and detail their own safety culture.

If a detailed formalization can help to describe the safety culture and help to assess it through surveys, several examples show that there is not one single safety culture in an organisation.

Indeed, in addition to present a formalized and homogenous component described in documents, a safety culture in an organisation can present several informal professional components which gather skills and representations, locally developed in situation. Moreover, these professional safety cultures can be different or divergent. So, improving safety culture implies combining together several safety cultures in an organisation and to coordinate these professional safety cultures with a more formalized and homogenous one. How can this be achieved?

Furthermore, how can an organisation coordinate a homogeneous model with local situations?

In conclusion, safety culture is embodied in formalized and homogeneous documents and in several and informal professional safety cultures. The challenge is not to favour one of these options or to confront one another but to articulate both of them. In fact, a formalized safety culture can be a global common language which should be appropriated by professions and their managers in order to describe their own safety cultures, or to focus on their main characteristics or weaknesses. After that, managers can try to articulate these professional cultures and the formal safety culture.

3.2 How to Apprehend the Safety Culture Notion?

3.2.1 Safety Culture: What for, and for What?

Where is safety culture located in the risk management landscape from an industrial point of view? Is it a medium of a virtuous transformation of the organisation, a brick amongst others, necessary for good safety, just a tool of characterisation of a human group that is useless for an industrial, or something else?

For the industrial world, safety culture has become a subject of reflexion, of debate, of action and … of internal and external communication. Is the awaited effect on safety in line with this spending of resources and energy?

---

[9]Nuclear Regulatory Commission.

Another question arises rapidly. Which "safety" are we speaking about: safety of the industrial process or safety of the workers? Will strengthening the safety culture have the same impact on industrial safety and on occupational safety?

In fact, the distinction that has to be made is perhaps between minor and frequent accidents with simple scenarios on the one hand, and serious and rare accidents with complex scenarios, on the other. The level of integration of these two types of safety, the combination of policies and the choice of methods must take into account the benefits, the results and the objectives of both.

3.2.2 Safety Culture in Projects and International Aspects

Construction projects in industries are usually not located in the original country of the main company managing the project. Many different contractors can therefore be involved with employees coming from many different countries. According to the high level of risks related to some project activities, a strong Safety Culture or a Safety Model is expected from the stakeholders and the shareholders in a short time, because projects have to be successfully implemented within a few years.

In this respect, the objective of project management team is to reach a high level of safety performance right from the beginning of the project. In that sense, is there any existing specific Safety Model or Safety Culture program for projects?

If so, can industrial companies use the same method in different countries? As Safety Culture is influenced by components like location, company culture or the metier, which culture(s) predominate: trade, country, company? Is privileging one of them a good way to improve the Safety Culture? If yes which one?

Eventually, how can the notion of safety culture of the main company managing the project be extended to suppliers and subcontractors? And what about our own activities as contractors or shareholders for other partners with different kinds of culture?

4 Conclusion

As part of the FonCSI "strategic analysis" group on safety models and safety culture, representatives from four high risk-industries (EDF, SNCF, ENGIE and TOTAL) have shared their common issues and questions in the field. Considering their respective contexts of high industrial risks, both in terms of process safety and occupational safety, a certain number of topics arose for discussion, presenting so many challenges for the academic world.

Firstly, many safety theories, concepts, or models coexist today, which are more or less appealing and/or directly useful to the industry. How to choose from the available panel? And based on which criteria? For example, several safety approaches or models exist: what exactly do they comprise? What is the approach to take according to a given context (and *is there* one best way?)?

Should a unique model be considered, or several? Between a homogeneous safety model and a multiplicity of local models, what is the best option? What more does this notion bring?

Above all, the specific topic of "safety culture" emerges: how does this concept position itself with regards to other theories or concepts in the "safety cloud"? Can it link safety at the workplace and technological risks? And, again, how do specific local contexts and cultures influence safety outcomes?

So many questions left in suspension, the answers of which would shed light on operational decisions and safety strategies in the industrial world. Academics and Consultants, our friends, the floor is yours: make the sun emerge from behind the (safety) clouds!

5 Disclaimer

The views and opinions expressed in this chapter are the sole responsibility of the authors and may not reflect those of the companies they work for.

References

Guldenmund, F. W. (2000). The nature on safety culture. *Safety Science*, 215–257.
INSAG. (1986). *Summary report on the post-accident review meeting on the Chernobyl accident*. Vienna: IAEA.
INSAG. (1988). *Basic safety principles for nuclear power plants*. Vienna: IAEA.
INSAG. (1991). *Safety culture*. Vienna: IAEA.
INSAG. (1999). *Management of operational safety in nuclear power plants*. Vienna: IAEA.

Chapter 2
Safety Models, Safety Cultures: What Link?

An Introduction

Claude Gilbert

Abstract In this introductive chapter, Claude Gilbert, President of the FonCSI "strategic analysis" group on safety models and safety culture, shares with us the group's initial findings on this topic. This text was also used as the introduction to the research seminar organised in June 2016, key step of the project that led to this book. Depending on what is meant by "model", the way to address the link between safety models and safety culture will be different, ranging from straightforward to very complex. This chapter adopts a viewpoint focused on the actors concerned by safety, and questions how they are led to "navigate" through a world of constraints and opportunities. It highlights the importance to consider what is "already there" (company cultures), what may drive the organisations choices among the multiple offers available on the "safety culture ideas' and methods' markets". It ends up giving food for debate by proposing some research avenues to help industrial organisations better meet their expectations in terms of safety culture.

Keywords Safety models · Company culture · Constraints · Trade-offs

1 A Simple Question?

FonCSI's industrial partners asked a question to the FonCSI "strategic analysis" group, addressing the link between safety models and the safety culture in order to increase safety within companies carrying out hazardous activities. Depending on the approach, this question can be very straightforward or, on the contrary, become rather complex. Furthermore, it assumes a consensus on the very definition of safety, which is far from being the case.

The question is straightforward when the "model" is considered as a prescription, in the sense of a "model to follow" and something that can contribute to improving safety culture. Safety culture is then often associated with (a) an

C. Gilbert (✉)
CNRS/FonCSI, Grenoble, France
e-mail: claude.gilbert@msh-alpes.fr

C. Gilbert et al. (eds.), *Safety Cultures, Safety Models*,
SpringerBriefs in Safety Management,
https://doi.org/10.1007/978-3-319-95129-4_2

awareness of hazards and risks and (b) the way people involved in these activities adopt individual or group behaviours in order to manage these risks in the best possible way. In this approach, one can distinguish between two elements: a production of knowledge accompanied by recommendations and procedures, and conscious human beings whose behaviour must comply to these recommendations and procedures. This is a classic scenario. It relies on the domination of knowledge over action and on the distinct roles of those who think (researchers, consultant, etc.) and those in action in the field. In most cases, the expectations of actors who want better safety (e.g. industrial companies; regulatory authorities) are expressed in reference to this scenario.

The question becomes less straightforward when the "model" is considered from a more analytical rather than prescriptive perspective. It then refers essentially to the work carried out by researchers and consultants. This work distinguishes between various configurations of reference frameworks, organisational structures and practices that are typical of an industry (Amalberti, 2013) or even, of a company. Although this type of work can lead to recommendations and procedures, that is not its primary goal; even less so given models (resilient, safe or ultra-safe, for example) prevail in the industries and companies concerned for survival reasons. These reasons differ from one industry to the next, as does the relationship to safety: civil aviation and the nuclear sector have no choice but to have ultra-safe safety models, just as sailors and fishermen have no other choice than to adopt a model that is simply resilient. When we approach the "model" from an analytical angle, the question arises as to how important safety is or is not, considering the various constraints to which the different industries are subjected. Moreover, such an approach partly blurs the boundary between safety models and safety culture. Indeed, practices that are inseparable from what is usually meant by safety culture are used to define safety models.

The question initially put forward is no longer straightforward at all when we do not consider safety models to be external tools that can/must be applied to individuals and work groups in order to improve safety. This is the case when we take as a starting point to the analysis the safety culture as it is already present within the different industries and companies, and thus as we can understand it from an anthropological perspective. In other words, when through the notion of culture we seek to embrace what certain sociologists call the "already there" (Lascoumes, 1994), i.e. everything that existed prior to the desire to make safety-related changes. Thinking from this perspective means firstly considering that within industries and companies there are already safety cultures that simultaneously incorporate "safety models" (with a coexistence of old and new models), local knowledge (sometimes formalised, sometimes not), and know-how that results from practical experience (with various ways of sharing and transferring it). The whole corresponding to a quite baroque combination. It also means considering that these safety cultures—combinations of distinct elements—correspond to the trade-offs made between the various constraints to which the industries and companies are subjected (profitability, business continuity, safety, preservation of social harmony, etc.). When, instead of throwing ourselves head-on into the pursuit of the desired future and making a clean sweep, we first try to find out how what is “already there” is configured, the perspective changes. When it comes to safety, the challenge lies in focusing on effective possibilities for change given what already exists, while also taking into account the internal and external contexts within which these changes are to take place.

2 Shifting the Question

The discussions that took place within the FonCSI group establish a link between these general or even abstract questions, while also shifting the questioning. Rather than questioning the safety models/safety culture combination (in one direction or the other), it would seem preferable to “situate” the actors concerned by safety (in the first instance, those in companies), seeking to promote it in relation to a set of constraints and opportunities (Fig. 1).

In fact, industrial companies are at the conjunction of:

“Company cultures”, which correspond to what is “already there”, reflecting everything that has been established and accepted in the way of looking at and implementing safety (via the general activities) and which has been incorporated into the organisational structures, procedures, habits of individuals and work groups, technological choices… through which safety is truly embodied, “the dead seizing the living” (*«le mort saisit le vif»*) to quote an expression used by Pierre Bourdieu (1980). Cultures buried within the actual reality of companies and that

Fig. 1 The safety models and safety cultures triangle by Hervé Laroche and members of the FonCSI “strategic analysis” group

rather imperceptibly carry a great deal of weight in that they determine both the possibilities and impossibilities for change.

A safety-related offer in the ideas market fuelled by intellectual output (concepts, theories, methods) from the academic world and from consultants, and which can have analytical or prescriptive aims. An offer which, as it spreads across both the academic field and the field of consulting, sometimes spilling over into the public arena, can incite action or a manifestation of the intention to act. Indeed, it is regularly in reference to these potential resources that public debates arise when incidents, accidents or crises occur.

A set of safety-related approaches that aim to improve safety within companies, in any form whatsoever. These approaches can have internal origins, since different categories of actors can, depending on the circumstances, "have an interest" in promoting safety. Some research has indeed highlighted the fact that safety, which is a cross-functional issue within companies, can be a lever for different types of action or even a way to gain power (Steyer, 2013). The approaches can also have external origins and stem, for example, from actors that, for various reasons, are looking to demonstrate specific skills in the area of safety and thus position themselves in what is in fact a market (connected, of course, to the ideas market).

Therefore, the choices made in relation to safety, whether these are to promote models and/or strengthen cultures, will not be based solely on rational acts (in the generally understood sense). They will also result from the way in which industrial companies are led to "navigate" through this world of constraints and opportunities, through this "force field" (Chateauraynaud, 2011), once they are required to act (whether due to a deliberate desire to achieve efficiency or whether in reaction to requirements or requests for justification from the environment within which these companies operate—regulatory authorities, media, civil society, etc.).

A few lessons can be drawn from this:

It is probably unrealistic to believe that "good" safety models exist per se, either as a result of academic or expertise work, or even as can be evaluated internally based on efficiency criteria. A good safety model is indeed a model that fulfils objective requirements, which specialists are skilled at setting, but also a model that is compatible with the culture of the company in question or, at the very least, that provides levers to understand the company such as it is configured by the culture that characterises it; a model that gives "good reasons" to act and that provides elements to justify this (particularly with regards to external actors acting as observers or even critics).

It is probably unrealistic to believe that it is possible to strengthen the safety culture solely by disseminating "good" safety models, particularly if these focus on raising the awareness and influencing the behaviour of individuals (as it is still partly the case today, despite the emphasis placed on the organisational aspect). Whether or not they involve the integration of new safety models, actions that are effective and long-lasting in this area are those that get to the very foundations of the company culture, "buried" as it is in the company's procedures, organisational structures and practices, and those that successfully re-open the "black boxes" that have thus formed.

3 So What?

Based on these observations, what are the avenues to explore in order to meet the expectations of industrial companies, even if it means shifting the questioning (as is actually expected from the FonCSI group)?

First avenue: emphasise the need to truly establish what is "already there" before engaging in any deliberate action in the area of safety. This is a difficult task for several reasons. Going down this path means recognising that the reality of companies is particularly complex with the accumulation and layering of different technologies and organisational options that correspond to different logics. It also means recognising that even though safety can correspond to specific competencies, it is largely diluted within the general activities of companies and is thus part of their general culture. Consequently, understanding what already exists and its many consequences requires investigation and analysis work that can seem a priori costly (not only financially, but also in terms of time, investment, efforts spent elaborating specific diagnostic tools, even if it is based on recommendations from the IAEA[1]). It can also be costly on another level, since it can produce an image of companies that does not match the one presented for marketing and communications purposes. Nevertheless, this cost should be compared to that of not proceeding down this path (and all it entails).

Second avenue: consider that safety models, as they are introduced in the academic and expertise market, are resources for the internal and external actors who, for various reasons, act as the promoters of safety. Going down this path means recognising that, beyond the fact that it seems to be an evident problem that needs solving, safety represents a challenge but also leverage for power within and outside of companies and it can act as a "springboard". Therefore, it means recognising that the different advocates of this cause are driven by different types of interests and that they can potentially be used to promote safety. In this case, it should be determined which of these actors and groups of actors are, in given circumstances, the most able to do what is required. The difficulty lies in the fact that any decisions are then as scientific as they are technical and political.

Third avenue: consider safety-related actions as being part of a company's strategy and not simply as the application of procedures. Going down this path means recognising that safety is not a technical matter and that it is above all a strategic matter, given the existence of tension or even contradictions between the safety culture as it is incorporated into the general company culture and safety as it is presented in models. It therefore means recognising the need to find the actors and action plans that are most likely to find ways to create interfaces between the "existing", or what is "already there", and the planned changes contained in the models. In particular, so that these changes are long lasting. The difficulty, then, lies

[1]International Atomic Energy Agency.

in identifying the "go-betweens" who have enough tactical or even political sense to make these changes or to guide them, and to allocate the necessary resources to their implementation.

All of these avenues and the perspective from which they have been formulated will no doubt give rise to much debate. Questions have already emerged within the group following the analysis it carried out: "If safety can be alternately an object of power, a strategic company objective or an academic subject, what is its essence?" Or, "If safety is the subject of exchanges between authorities that develop regulatory requirements, actors that resist (or not) and experts that make recommendations, what then is the nature of the exchange that takes place?" And, from an even broader perspective, "What is the economy of this ecosystem?" Indeed, "Though the company is at a crossroads, it is also at the centre of a hub…". In closing, it is clear that the approach suggested by the FonCSI "strategic analysis" group raises new questions even though it is already possible to identify concrete avenues for action.

References

Amalberti, R. (2013). *Navigating safety*. New York, NY: Springer.

Bourdieu, P. (1980). Le mort saisit le vif [Les relations entre l'histoire réifiée et l'histoire incorporée]. *Actes de la recherche en sciences sociales, 32,* 3–14.

Chateauraynaud, F. (2011). *Argumenter dans un champ de forces. Essai de balistique sociologique*. Paris: Editions Petra.

Lascoumes, P. (1994). *L'éco-pouvoir, Environnements et politiques*. Paris: La Découverte.

Steyer, V. (2013). *Les processus de sensemaking en situation d'alerte, entre construction sociale du risque et relation d'accountability, Le cas des entreprises françaises face à la pandémie grippale de 2009. Thèse de doctorat en science de gestion*. Paris: Université Paris Ouest-ESCP-Europe.

Chapter 3
Understanding Safety Culture Through Models and Metaphors

Frank W. Guldenmund

Abstract "*Few things are so sought after and yet so little understood.*" With this pithy statement, psychologist James Reason expressed the potential value but also the elusiveness of this complex social-scientific concept twenty years ago (Reason, Managing the risks of organizational accidents. Ashgate, Aldershot, 1997). Culture had been on the mind of safety scientists since Turner's book *Man-made disasters* from 1978, but the term 'safety culture' was only coined nine years later, right after the Chernobyl nuclear disaster in 1986. Since then, safety culture has been alluring as a cause—for both occupational accidents and process related events—and as a thing to strive for, although possibly unattainable (Guldenmund, Understanding and exploring safety culture. BOXPress, Oisterwijk, 2010). In this chapter, I will look at various perspectives on (safety) culture, using the metaphor as an illuminative principle, to identify (what seems to be) the essence of some dominating perspectives. Firstly, however, a common understanding of what culture 'is', needs to be established. I will then touch upon the assessment of culture. Afterwards, I will present four metaphors for safety culture, which represent the dominant perspectives on this concept. The chapter ends with suggestions on how safety culture might be influenced.

Keywords Safety culture · Culture model · Culture development
Culture assessment · Culture metaphors

1 Understanding Culture: A Brief Introduction

What *is* culture? Culture emerges at places where people live and work together. Living and working together requires a certain degree of shared understanding—e.g. about daily reality, about work and its context, and so on—and it is this

F. W. Guldenmund (✉)
Delft University of Technology, Safety & Security Science Group, Delft, The Netherlands
e-mail: f.w.guldenmund@tudelft.nl

C. Gilbert et al. (eds.), *Safety Cultures, Safety Models*,
SpringerBriefs in Safety Management,
https://doi.org/10.1007/978-3-319-95129-4_3

(shared) understanding that a culture provides. Culture exists *between* people[1] and is activated when they meet, see symbols from, or perform rituals pertaining to a culture they have adopted. Different people and different contexts evoke different cultures and people usually carry several different cultures. The actual essence of culture is notoriously hard to define, but it embodies values, norms, meanings, convictions, beliefs, assumptions, and so on, that enable people to make sense of their world and perform in it as well as to make sense of other people's behaviors. Their behavior could be words or deeds, but could also consist of strong feelings or opinions that are either articulated or expressed more implicitly.

Culture plays a crucial role in society and, consequently, in organizations. Culture influences, but is also influenced by, the structure and formal part of an organization as well as the daily execution of its processes. The latter occurs as interactions between people and between people and the primary process. If safety is an integral part of this primary process, the resulting culture is called 'safety culture'.

1.1 Definitions

There are many definitions of culture and they often overlap considerably (Antonsen, 2009). Three definitions express the description of culture given above well. Culture is

> the collective programming of the mind which distinguishes the members of one group or category of people from another. (Hofstede, 1991, p. 5)

Hofstede's definition highlights that culture is acquired, as 'mental software', as well as the distinctive nature of cultures, between groups or categories of people.

> Culture is a fuzzy set of basic assumptions and values, orientations to life, beliefs, policies, procedures and behavioural conventions that are shared by a group of people, and that influence (but do not determine) each member's behaviour and his/her interpretations of the 'meaning' of other people's behaviour. (Spencer-Oatey, 2000, p. 3)

Spencer-Oatey emphasizes the influence culture has on people's behavior as well as its interpretative aspect, through which people are able to understand each other's behaviors. Based on this understanding people can make attributions as to why people do the things they do.[2]

Finally, the Norwegian Bang defines (organizational) culture as

> the set of common norms, values and world views that develop in an organization when its members interact with each other and its context. (Bang, 1995, cited in Martinussen & Hunter, 2018)

[1]Although, of course, it is ultimately coded in their brains.

[2]Which might, nevertheless, turn out to be false, as one group's behaviors or symbols might have a different meaning than another group's.

Bang's definition describes the interactive and context-dependent nature of culture. Culture emerges spontaneously, unintentionally even, whether a group likes it or not. On the one hand, this highlights a certain arbitrariness of its content and the fact that it might have turned out differently with a different group interacting under different circumstances. However, when a culture has established itself within a group, its members are cautious and often even unwilling to adapt it, unless the group becomes ineffective or dysfunctional (Schein, 2010). On the other hand, it is difficult to say who decides what becomes part of culture and what not. Again, this is entirely dependent on the composition of the group, its context and the task(s) at hand.

1.2 The Nature of Culture

It is hard not to write about culture as 'something' that a group 'possesses', as if culture is an instrument used by a group. Hofstede's allusion to the term 'software' might be particularly useful. Software as an operating system, provides the rules and procedures for the computer to 'behave', i.e. to compute. Software is not a thing as such, which can be manipulated by the computer to its own liking or benefit. Software can be updated or changed indefinitely, and the same goes for culture.[3] However, this is also where the comparison stops, as software can be changed and updated by command, whereas culture cannot. Comparable discussions on the nature of organizational culture were carried out in the 1980s when scholars queried whether organizations basically 'have' or 'are' a culture (e.g. Smircich, 1983), resulting in functional and interpretive paradigms (e.g. Guldenmund, 2016).

1.3 Schein's Culture Model

Schein's model of culture perhaps provides some grip on a culture's elusiveness and is, moreover, especially relevant for safety. Schein defines culture as a core consisting of implicit and covert basic assumptions (strongly held beliefs, values, norms, and so on) surrounded by two overt and, hence, empirically tangible, layers (Fig. 1). 'Artifacts', the outer layer, are clearly visible but not directly convertible to an underlying culture although they might be an expression of it (or not). Artifacts are most easily acquired and can function as a façade rather than a cultural expression. The same goes for 'espoused values', the second layer which, again, are tangible but do not need to translate directly into underlying basic assumptions. Espoused values are the values people express when asked about something.

[3]But through entirely different mechanisms and time scales.

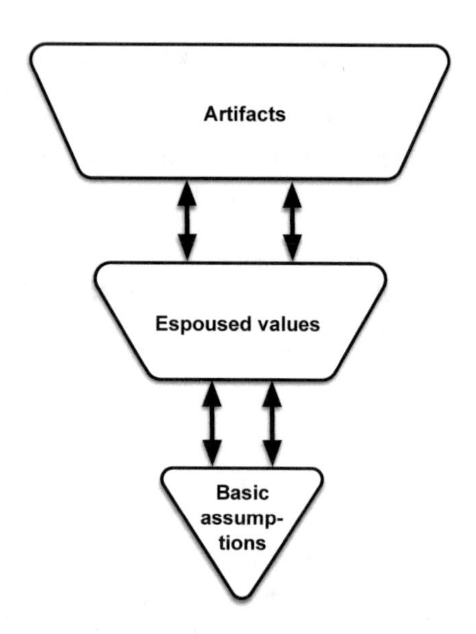

Fig. 1 Schein's model of culture

However, they might not reveal the actual culture lying underneath, as they may rather be (good) intentions, (future) ambitions, social desirable notions or politically correct answers (Banaji & Greenwald, 2013). Indeed, no organization will publicly espouse that they do not care about safety. While 'zero accidents' or 'safety is our first priority' might be their maxim, daily reality might nevertheless prove otherwise.

Schein's model typically provides an outsider's view on culture, and he warns this outsider not to take his (or her) observations at face value. (S)he needs to reach beyond the artifacts and espoused values to truly understand the underlying culture. This is a process of observation, interpretation and confirmation; a course of investigation defined by Schein as 'deciphering' (2010).

1.4 Culture Development Model

A further hold on culture can be provided by describing the process through which culture develops and is maintained within a group. The first step, called 'understanding' (Fig. 2), is an essential and individual activity people have to carry out in order to survive in the world, which does not speak for itself. People need to interpret what they see and make attributions based on these interpretations. This goes for objects and behaviors but especially for concepts—like hazard, safety and risk—which are, by definition, abstract. Of course, understanding does not start

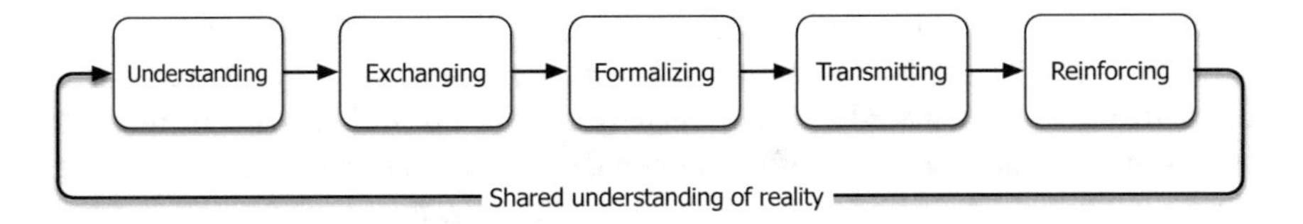

Fig. 2 The development of culture

from scratch. People bring an assembly of cultural assumptions and mental models to different (cultural) contexts. These contexts themselves are subject to cultural influences, notably regional or national, but also influences handed down through education, that is, professional cultures. Moreover, the cultural inputs received early in life reside deeper in one's cultural core than the inputs received later on (Hofstede, 1991). And what is deep in terms of culture, is less susceptible to change and is able to evoke strong emotional reactions when challenged or threatened (Ibid.).

So, a person's understanding of a particular situation at work is colored by assumptions coming from different cultural sources, all of which reside within the boundaries set by the deepest, most basic cultural notions acquired mostly in childhood. Nevertheless, an infinite number of meanings are still ambiguous and open to multiple interpretations; for instance, Weick refers to such instances as 'equivocality' (1969, 1979). Again, the topic of safety is relevant here, as safety is not self-evident and needs to be elaborated to make sense locally (Hollnagel, 2014). Such ambiguity is often resolved through the second step in the process, exchanging.

This exchange between people is preferably based on dialectics (dialogue), which builds on an open exchange of the participants' viewpoints and ideas. Ideally, this results in a common understanding of the situation as well as an action to make further progress possible. This selection of an interpretation plus action is then taken as the right way to think and work if it indeed resolves equivocality and enables progress of work. As this understanding proves itself in practice, it is retained, perhaps fine-tuned further and, subsequently, formalized. However, the exchange is more often *not* based on dialectics. Bringing their own assumptions and convictions to the workplace, people will try to convince others of their viewpoint and be unwilling to exchange it for the others'. This results in opposition and even conflict. This could be resolved with agreement, "Let's agree to disagree", but it depends on the importance of the issue, the coherence of the group and how this conflict will impact on the group and its work. Either an agreement or a conflict might be part of a culture. Conflict ultimately marks the boundaries of what is acceptable for the group, whereas agreement binds the group. Although some disagreement or conflict within a culture is desirable for 'requisite variety'—the variety necessary to cope with a constantly changing environment (Ashby, 1958)—, too much conflict is unwelcome (Alvesson, 2012) since it will threaten the group's coherence and might lead to fragmentation and, ultimately, its collapse.

The process of culture development implies that both formal and informal structures or standards follow from interaction and subsequent agreement. Formalization and standardization ultimately lead to institutionalization, implying that norms and standards are enforced upon the (working) environment where they can be put into practice and complied with. In this way, the (working) environment is both a source for and a reflection of a culture's underlying assumptions and meanings.

Furthermore, what is institutionalized will be transmitted and disseminated. Newcomers need to learn the rules, whether explicitly, through education and training, or implicitly through interaction with current group members, a process called enculturation. As the classic experiments of Asch show, people indeed conform to group behaviors easily, even when they disagree or think they are downright silly (Zimbardo, Johnson, & McCann, 2012). When cultural meanings are stressed again and again, through interaction, through reinforcement, through working in a context that consistently expresses and confirms those meanings, they will gradually become part of the individual's cultural core, unless (s)he opposes to them. Armed with these cultural meanings, large parts of the world appear unambiguous whereas in reality, they are not.

1.5 Culture Integration

Cultures are neither homogeneous nor fully integrated. On the one hand, because disagreement and even conflict will always arise. On the other hand, because people within a culture adopt its core with mixed intensity. What is a guiding principle for one, might be a hollow cliché for another. A large group therefore always displays differentiation and consists of various subcultures.

1.6 Elaborating the Development Model

While the development model above describes a rather cumbersome process, it is not always initiated in full. When business is 'as usual', the local culture provides interpretations and courses for actions for those involved. This basically means that the arrow coming out of the first box (Understanding) in Fig. 2 immediately inputs into the fifth box (Reinforcing). In other words, when somebody acts in line with the local culture, their assumptions are, hence, reinforced. However, when workers are confronted with equivocality, i.e. when something in the present cannot be understood with current (cultural) assumptions, or otherwise, a process Weick calls 'sensemaking' is initiated, an exploration seeking to answer the question 'What is going on here?' (Weick, 1995). This is when the full process of Fig. 2 starts. While equivocality might always be present in the (working) environment, people also have a tendency to ignore or not even notice this and look for confirmation for what they already assume; a phenomenon called 'confirmation bias' (Nickerson, 1998).

Newcomers at times, will have trouble understanding their new (working) environment. They will be either informed locally, or educated or trained, which, in the latter case, means they start at Step 4 in Fig. 2 and are exposed to more formal transmissions (education, training).

2 Safety Culture Revisited: Images of Culture

Currently, people interested in safety culture do not always adhere to a single view. Instead views range from quite abstract (conceptual) to concrete (instrumental), from straightforward to complex, from (richly) descriptive to analytical. Theoretical developments that complement each of these particular views on organizational safety culture differ, as do the accompanying research methods. What follows is a brief overview of these perspectives using several images. They are not mutually exclusive and may overlap. Some images are more popular than others, however, clinging to a single perspective is neither recommended nor considered fruitful.

2.1 Safety Culture as a Convenient Truth

There are convenient and inconvenient truths. An inconvenient truth can be a cause for embarrassment, a convenient one a cause for relief.

There is still keen interest in safety culture, especially with safety professionals and high-risk organizations. Often, for convenience's sake, safety culture is equated with (un)safe behaviors, or daily practices. But whose culture is implied here, and whose behaviors? It depends, of course, on who is talking. When it is management, the culture and behaviors of front-line workers are insinuated, as in "*They* don't do what they've been told, *they* make up their own rules". When it is the workers, a them-or-us mentality might be in place, as in "*They* don't know what work is really like at the front end". These are insider or first-person perspectives, provided from either 'above' or 'below' and are used to shift misunderstanding, from one end to the other.

An outsider or third-person perspective is offered by people in various roles:

- evaluators: inspectors, regulators, auditors, insurance companies;
- investigators: investigation boards or committees, accident investigators;
- advisors: consultants, change agents;
- other stakeholders, like investors.

When outsiders refer to culture they imply the (dominant) organizational culture, as something the organization reflects, or even 'is'. This culture is for them the cause of what is currently happening or has recently happened, like an incident or accident, in the organization. And it is the outsider perspective that might use the safety culture label as a convenient truth, perhaps to appoint no blame at all.

Talking about the (safety) culture of organizations in this way is convenient and easy but not necessarily truthful or helpful. Understanding how an organization understands (and sustains) its daily reality and the role of safety, and how this might influence the behavior of its employees is less easy and convenient, but perhaps more truthful.

2.2 *Safety Culture as a Grading System*

Probably the most dominant reason for assessing safety culture is to grade it. Keywords in this metaphor are measurement, benchmark and, if possible, improvement.

So, what is measurement, exactly? Measurement is

> assigning numbers to the values of a natural variable in such a way that relationships between those values turn into similar relationships between the numbers assigned. (Swanborn, 1987)

Measurement scales satisfy one or more properties of measurement.

1. Each value on the scale has a unique **identity** or meaning.
2. Values on the scale have some **magnitude** and an ordered relationship to each other, i.e. some values are smaller and some are larger.
3. The **intervals** or units of the scale are all equal to one another.
4. The scale has an **absolute zero point** below which no true values (can) exist.

The nominal measurement scale satisfies the identity property only, the ordinal scale satisfies both the identity and magnitude properties whereas the interval scale also has equal intervals. There are no culture measurement approaches satisfying all four properties, the so-called ratio (or cardinal) scale is not common in the social sciences at all (unless frequencies are used). All three measurements scales are represented in the safety culture assessment toolbox, but grading only starts at the ordinal level.

The 'natural variable' referred to above is safety culture. When using a nominal scale for this purpose, it would simply mean putting different labels on different types of safety cultures. For instance, Cameron and Quinn put forward four (organizational) culture types—clan, adhocracy, hierarchy, market—which group particular characteristics typically found in organizations (Cameron & Quinn, 2011). However, such culture types are not usually ordered, with one 'better' than the other. But that is precisely what a *grading* system needs to establish, what is sufficiently good as well as opportunities for improvement. Moreover, a grading system in theory makes comparisons possible. And benchmarking is attractive from a management point of view. Hence, grading systems at the ordinal level of measurement or higher are called for.

Safety culture assessment techniques at the ordinal level of measurement are usually represented by ladders; i.e., stairways leading to a desirable level of safety performance. Progression on the ladder is established by growth in maturity in dealing with the safety issues addressed by the method.

The general appeal of these culture ladders is high in some circles, yet they still lack proper scientific underpinning. However, what is the use of 'scientifically sound' methods when dealing with something as elusive as '(safety) culture'? Moreover, these methods are not meant to measure safety culture in the first place but to carry out a meaningful dialogue amongst employees about safety. The grading system only serves as a means to challenge people, to reflect on and take position on various safety issues. The grading system used in the Hearts and Minds toolbox, indeed uses some provocative labels. It runs from pathological, reactive, calculative, proactive to generative (Energy Institute, 2016). Because of their disapproving undertone, the first three labels might stimulate discussion amongst a group of people, if they are willing to challenge each other. Used by a third party in order to grade companies, the method loses much of its initial appeal and it becomes just another way to say you are 'good', or 'bad' (or 'adequate').

Safety culture assessments have also climbed up the measurement hierarchy,[4] which brings us to the interval level of measurement. In the social sciences, they are usually carried out with standardized questionnaires. What is measured here is more accurately referred to as 'safety climate', considered by Zohar (and many others) as 'the measurable aspect of safety culture' (Zohar, 2014). Indeed, a safety climate survey enables researchers to put exact numbers on (aspects of) the safety climate, reported sometimes with an accuracy of two or even three decimals. However, such numbers are mostly used as descriptors rather than grades.

2.3 *Safety Culture as a Liaison*

Keywords in this metaphor are operationalization and standardization. Operationalizing is the act of making an abstract concept empirically tangible, often by breaking it down into manageable, i.e. tangible or measurable, parts. The concept here is safety culture; the liaison is the connections between the constituting parts. The parts are usually concepts themselves, but these have been derived empirically, again with the use of standardized, self-administered questionnaires. Responses to these questionnaires go through an analysis process aimed at uncovering (or confirming) an underlying structure. This structure consists of the

[4]Safety culture assessment approaches can be ordered on a measurement continuum as well; that is, there is a nominal, an ordinal and an interval-level approach. The latter two are *etic* approaches, the first one can be considered *emic* (see Sect. 3.1).

measurable parts mentioned above. For instance, the concept of safety culture is regularly broken down into parts that measure the workforce's perceptions of the ways the organization deals with safety.

Note that this way of describing safety culture has been labelled safety climate above. Safety climate, as opposed to safety culture, is a 'psychological variable', describing attitudes and perceptions typically assessed at the level of an individual employee. These scores are aggregated to some group level (team, department, organization), and their homogeneity is determined to see whether the group's perceptions are unanimous, or not, on the various safety climate aspects. As opposed to the grading system, the measurements obtained here are used descriptively and statistical tests are carried out to examine any differences between existing groups.

Standardization is part of the metaphor as well. Safety climate is considered to be a widespread phenomenon that can be tapped by using the same questionnaire everywhere. This notion is also called 'nomothetic', implying that the underlying structure of the safety climate questionnaire is relevant to working people all around the globe. From another perspective, however, this approach could be perceived as forcing respondents into the researcher's theoretical framework. Moreover, as questionnaires are devoid of context, the desktop researcher does not have a clue why the respondents answered the way they did.

2.4 *Safety Culture as a Mirror*

The image of the mirror evokes the act of measuring up and scrutinizing oneself. There are different mirrors for different purposes and the mirror can be held by various people for different reasons. In the context of safety however, this act should be (but frequently is not) carried out critically and not admiringly. Organizations often want to know where they stand, whether there is room for improvement and what has to be done to succeed. Looking into a (cultural) mirror can provide some answers.

The mirror held up by an outsider can be any of a range of safety culture studies, either ideographic (say, a tailor-made, qualitative field study) or nomothetic (using standardized methods). Let us focus on the introspective study of organizations looking into the cultural mirror themselves. The organization thus describes its own culture, preferably using multiple methods, as no method alone can provide a full mirror image. Ideally, it works with an assessment team, composed from members selected from all layers of the organization. It is supported by top-management, which keeps itself updated about the progress of the team. Through the application of various methods, the team gradually peels off the cultural layers of the organization and works towards the core, the shared understandings, yet it does not ignore conflict or differentiation within the organization. Steadily working towards a mirror image, the organization starts to know itself, its shared beliefs and assumptions, its behavioral patterns and what sustains them. Such insight may also yield opportunities to influence, the final topic of this chapter.

3 Assessing and Influencing Culture

3.1 Assessing Culture

Cultures are described or assessed in most cases to influence it.[5] There are basically two perspectives on describing (or assessing) cultures, denoted by the terms *emic* and *etic*. Taking an *etic* perspective means applying a theory-driven, top-down approach using a standardized instrument, often developed through the application of the empirical cycle of positivism; i.e. observation, induction, deduction or prediction, testing and evaluation (De Groot, 1961). This standardized instrument is most often a questionnaire, an operationalization of the underlying theory. It contains dimensions (factors, aspects, facets) on each of which the culture under study is evaluated. Questionnaires appeared also in descriptions of the metaphors above, especially in the grading system and the liaison.

Using an *emic*-approach, a culture is evaluated on its own merits. The purpose is a description of what truly matters in this particular culture, how members understand their reality. Working from the bottom up using members' perspectives and words, the researcher comes to understand what the culture is about and why. This is a demanding process which requires the researcher to remain descriptive, not evaluative. Given the fact that all people come equipped with their own cultural cores, researchers using an emic-approach are well-advised to make explicit any second thoughts or misgivings they have about their enterprise before they embark on their study. In this way, they show their biases and how they might have slipped into their study and descriptions.

Emic-studies are often considered 'subjective' whereas etic ones are believed to yield 'objective' results. In the end, all studies are 'subjective'; however, standardization of an approach enables replication to the extent that another researcher using the same instrument will produce comparable results. Replication is not the objective of emic-studies and neither is objectivity. As long as the researcher carefully registers her or his data, another researcher can draw his (or her) own conclusions about these data.

[5]This is not true for anthropologists and some sociologists, who describe cultures for their own sake.

In the case of safety culture, an interesting clash occurs. Whereas the concept of culture is value-free (there are basically no 'good' or 'bad' cultures), the concept of safety is not: situations can be 'unsafe' and this is considered 'bad'. Of course, to establish that a situation is unsafe one needs some kind of norm and there are many of these available. As already seen, this clash of concepts has been resolved in different ways, resulting in different metaphors.

3.2 Influencing Culture

As culture development is a constant and continuous process, any culture is subject to ongoing change or adaptation, especially when the very existence of the group is threatened (Schein, 2010). Because culture is the result of a rather time-consuming process, opting to influence it implies another demanding effort (Ibid.). So, a first consideration should be to use the particular strengths of a culture to achieve desired (safety) objectives. Nevertheless, tweaking parts of a culture might turn out to be more opportune.

Influencing culture starts with influencing the way employees understand their working reality and the place of safety therein ('Understanding', see Fig. 2). The type of risk communication employing rhetoric, sometimes also referred to as 'care communication', is concerned with influencing people's understandings of and attitudes towards risk (Lundgren & McMakin, 2013). Moving to the second step of the culture development process, 'Exchanging', it is not individual employees that need to be influenced here, but the way they interact and communicate about safety. In this step, the dialogue is a key mechanism. It is a way of obtaining consensus, of reaching a shared understanding of daily working reality, that works in practice too.

Other methods can be applied to influence the model's further steps. For each step, it is important that the result adds to a shared understanding (of daily reality), rather than providing a particular viewpoint that is not recognized by the group at large.

4 Conclusion

The popularity of safety culture has led to a labyrinth of papers and approaches on what it is and is not and how to assess it, mimicking a similar boom in the 80s on the topic of organizational culture. Unfortunately, this lack of consensus also reflects negatively on the safety culture concept itself, as it has become a true 'God-of-the gaps', to explain missing links in our understanding of safe and unsafe behavior.

Current safety culture approaches can be described using four metaphors; i.e. a convenient truth, a grading system, a liaison and a mirror. Apart from the first metaphor, these approaches are mainly concerned with the assessment of safety culture. Next to these, a development model has been put forward, which is less concerned with an outcome (what safety culture should be) but rather with how culture originates, i.e. how a group comes to understand its context and how to act (safely) in it. Each step in the model can be used to develop interventions to steer its outcome towards a more desirable, safer direction.

References

Alvesson, M. (2012). *Understanding organizational culture* (2nd ed.). London: SAGE Publications Ltd.

Antonsen, S. (2009). *Safety culture: Theory, method and improvement*. Farnham, Surrey, UK: Ashgate.

Ashby, W. R. (1958). Requisite variety and its implications for the control of complex systems. *Cybernetica, 1*(2), 83–99.

Banaji, M. R., & Greenwald, A. G. (2013). *Blindspot: Hidden biases of good people*. New York: Delacorte Press.

Cameron, K. S., & Quinn, R. E. (2011). *Diagnosing and changing organizational culture based on the competing values framework* (3rd ed.). San Francisco, CA: Jossey-Bass.

De Groot, A. D. (1961). *Methodology: Foundations for research and thinking in the behavioural sciences*. The Hague: Mouton. (in Dutch).

Energy Institute. (2016). *The hearts and minds safety culture toolkit*. Retrieved October 29, 2016, from http://heartsandminds.energyinst.org/.

Guldenmund, F. W. (2010). *Understanding and exploring safety culture*. Oisterwijk: BOXPress.

Guldenmund, F. W. (2016). Organisational safety culture. In S. Clarke, T. Probst, F. W. Guldenmund, & J. Passmore (Eds.), *The Wiley-Blackwell handbook of the psychology of occupational health & safety* (pp. 437–458). Chichester, UK: Wiley.

Hofstede, G. R. (1991). *Cultures and organisations: Software of the mind*. London: McGraw-Hill.

Hollnagel, E. (2014). *Safety-I and safety-II. The past and future of safety management*. Farnham, UK: Ashgate Publishing Ltd.

Lundgren, R. E., & McMakin, A. H. (2013). *Risk communication. A handbook for communicating environmental, safety, and health risks*. Hoboken, NJ: Wiley.

Martinussen, M., & Hunter, D. R. (2018). Aviation psychology and human factors (2nd ed.). Boca Raton, FL: CRC Press.

Nickerson, R. S. (1998). Confirmation bias: A ubiquitous phenomenon in many guises. *Review of General Psychology, 2*(2), 175–222. https://doi.org/10.1037/1089-2680.2.2.175.

Reason, J. T. (1997). *Managing the risks of organizational accidents*. Aldershot: Ashgate.

Schein, E. H. (2010). *Organizational culture and leadership* (4th ed.). San Francisco: Jossey-Bass.

Smircich, L. (1983). Concepts of culture and organizational analysis. *Administrative Science Quarterly, 28,* 339–358.

Spencer-Oatey, H. (2000). *Culturally speaking: Managing rapport through talk across cultures*. London: Continuum.

Swanborn, P. G. (1987). *Methods of social-scientific research*. Meppel: Boom. (in Dutch).

Turner, B. (1978). *Man-made disasters* (2nd ed.). London: Wykeham.

Weick, K. E. (1969). *The social psychology of organizing*. Reading, MA: Addison-Wesley Publishing Company Inc.

Weick, K. E. (1979). *The social psychology of organizing* (2nd ed.). Reading, MA: Addison-Wesley Publishing Company Inc.
Weick, K. E. (1995). *Sensemaking in organizations*. London: Sage Publications Ltd.
Zimbardo, P. G., Johnson, R. L., & McCann, V. (2012). *Psychology. Core concepts* (7th ed.). Boston: Pearson.
Zohar, D. (2014). Personal communication.

Chapter 4
The Use and Abuse of "Culture"

Andrew Hopkins

Abstract *Culture* is a misunderstood and misused idea. In this chapter I advance seven clarifying theses. (1) Culture is a characteristic of a group, not an individual, and talk of culture must always specify the relevant group. (2) Organisations have it within their power to ensure that organisational culture over-rides national cultures. (3) The most useful definition of the culture of a collectivity is its set of collective practices—"the way we do things around here". (4) In the organisational context, it is usually better to use culture as a description of group behaviour, rather than as an explanation for individual behaviour. (5) Organisational cultures depend on the structures that organisations put in place to achieve desired outcomes. These structures reflect the priorities of top leaders. The priorities of leaders in turn may depend on factors outside the organisation, such as regulatory pressure and public opinion. (6) The distinction between emergent and managerialist views of culture is misleading. (7) The term *safety culture* is so confusing it should be abandoned.

Keywords Culture · Safety culture · Meaning of culture · Sources of culture

The terms culture and safety culture are fashionable in safety circles and in business. Culture is a basic concept with roots in the disciplines of anthropology and sociology, but safety culture is a Johnny-come-lately, having arrived on the scene only in the latter part of the 20th century.

Both ideas are widely misunderstood and misused. Many writers have made this point before me. To mention just one, Hale (2000) wrote an editorial for an issue of *Safety Science* in the year 2000, entitled "Culture's Confusions". There is no agreement about the use of these terms, he said, and "confusion reigns". More than a decade and a half later, nothing has changed.

An earlier version of this chapter appeared in my book: *Quiet Outrage—The Way of a Sociologist.* CCH, Sydney 2016.

A. Hopkins (✉)
Australian National University, Canberra, Australia
e-mail: andrew.hopkins@anu.edu.au

C. Gilbert et al. (eds.), *Safety Cultures, Safety Models*,
SpringerBriefs in Safety Management,
https://doi.org/10.1007/978-3-319-95129-4_4

This chapter will not be a comprehensive discussion of these concepts. Instead I have chosen to advance a number of *theses* about culture. This enables me to cover several contentious issues and take a position on each. The discussion of each of is necessarily brief and perhaps overly dogmatic, but my aim is to provide accessible summary statements. Most of the theses concern culture; only the last will deal specifically with safety culture.

Given the nature of this book my focus will be on organisational culture, rather than culture as a more general sociological/anthropological idea, but it cannot be exclusively so, because organisational culture sits with that more general context.

1 Is Culture a Characteristic of Individuals or Groups?

Those seeking culture change within organisations often see the task as changing the values and attitudes of the individuals in that organisation, "winning their hearts and minds", creating an appropriate "mindset". There is an implicit assumption here that culture is a characteristic of individuals. However, social scientists insist that culture is a characteristic of groups, not individuals. Organisations may have multiple cultures and cultures may overlap and fragment into subcultures, but always the discussion refers to the characteristics of groups and subgroups, not individuals. Thus, one should never talk about culture without specifying the group, for example national culture, organisational culture, culture of the work group. This simple rule resolves many quandaries. The culture of the work group is not necessarily the culture of the whole organisation, and so on.

The claim that culture is the characteristic of a group, not an individual, has important implications. Consider the following statements made by the safety advisor of one large company.

> Safety performance has been achieved through an unwavering commitment and dedication from all levels in the organisation to create a *safety culture* which is genuinely accepted by employees and contractors as one of their *primary core personal values*. (Hopkins, 2000: 74)

The aim, he went on, is to "create a *mindset* that no level of injury (not even first aid) is acceptable".

The company drew an interesting implication from this. Since safety is about a mindset, the individual must cultivate it 24 h a day. It cannot be exclusively about occupational safety but must include safety in the home. Hence the company's 24-h safety program. This is how the safety advisor expressed it:

> Real commitment to safety can't be 'turned on' at the entrance gate at the start of the day and left behind at the gate on the way home. Safety and well-being of fellow employees is extended beyond the workplace in this company. A true commitment to safe behaviour is developed by promoting safety as a full time (i.e. 24 hour) effort both on and off the job.

All this depends on the idea that culture is a matter of *individual* attitudes. However, if one takes the view that culture is a group property, it may well be the case that attitudes to safety change as one passes through the factory gate. The company attitude to safety is one thing, but the attitude of a recreational peer group may be quite different, giving rise to much greater risk-taking outside the gate than inside. Think for example of attitudes to risk-taking in some motor cycle groups or hang gliding clubs. In both these contexts the aim is often to operate near the limit, without going over the edge. Sometimes a limit is transgressed, possibly with fatal results. Clearly, the same individual may have quite different attitudes to risk depending on the currently relevant group (Mearns and Yule, 2009). What the company referred to above is seeking to do, without realising it, is change the culture of groups *outside* the workplace. This it is most unlikely to be able to do.

Thesis 1 Culture is a characteristic of a group, not an individual, and talk of culture must always specify the relevant group.

2 National Versus Organisational Cultures

Companies sometimes complain that national cultures over-ride the corporate culture they are trying to create. The re-insurance company, Swiss Re, did a famous study a few years ago in which it identified "regional" differences in the oil, gas and petrochemical industries (Zirngast, 2006). One specific dimension was attitude towards safety, depicted in Table 1.

There are problems with this study, not the least being the rather grab-bag nature of the regions. Nevertheless, this study is sometimes taken as evidence that national cultures tend to over-ride corporate cultures. Indeed, that is the conclusion of the study.

> Our observation is that the influence of the country on the operational hazard is stronger than the influence from corporate headquarters. For example, a European [owned] refinery in the USA is currently more like a US refinery than a European refinery. (Zirngast, 2006: 8)

However the study author goes on to say:

Table 1 Swiss-Re study: attitudes to safety by region

Region	USA, Canada, UK, Australia	Europe, Singapore, S-Korea, Japan, Saudi Arabia, Gulf States, Egypt	Russia, Former Soviet Union, Eastern Europe	S-America, Africa, Maghreb, other Middle East, rest of Asia
Attitude to safety	Compliance driven, focus personal safety, fear OSHA, EPA, HSE	Respectful towards workforce, often positive safety culture	Unthoughtful	Company specific, focus personal safety

> We are open to the suggestion that the implementation of a corporate identity is possible.

In other words, the authors do not believe that the patterns they observed are inevitable. If global companies are willing to devote the necessary resources, they may be able to implement a uniform corporate style, no matter what the region.

This is supported by an important empirical study that concludes

> More proximate influences such as perceived management commitment to safety and the efficacy of safety measures exert more impact on workforce behaviour and subsequent accident rates than fundamental national values. (Mearns and Yule, 2009)

Shell's experience with a Korean shipyard it contracted with to build several vessels, nicely illustrates this whole issue. Shell was concerned that the fatality rate in Korean shipyards was very high, which was potentially attributable to Korean national culture. But it did not fatalistically accept this situation. It decided to supervise its contract closely and to insist that, where fatalities occurred, shipyard managers be dismissed. This policy was implemented and yielded dramatic improvements in safety. This demonstrates that companies are not at the mercy of local cultures and local ways of doing things. As the saying goes, where there's a will, there's a way.

Thesis 2 Organisations have it within their power to ensure that organisational culture over-rides national cultures.

If an organisational sets out to change its culture, how long will this take? I have heard consultants say that it can take five to seven years. The implication of the Korean shipyard story is far less depressing. As soon as there are real consequences for managers, cultures begin to change.

3 A Definition of Culture

There are many definitions of culture. Some attempt to be comprehensive and include so many components that they lack focus. But if we try to extract the essence of these definitions we find interesting differences. Anthropologists tend to focus on collective meanings. In contrast, in the context of organisations, definitions of culture tend to emphasise either values, or practices. The approach used by the company safety advisor mentioned above stressed values. The alternative is to emphasise collective practices: "the way we do things around here". The first thing to note about this latter formulation is the phrase "around here". Although vague, it makes clear that this is the culture of some group, perhaps a work group, or a larger organisational group. Second, the practices are inherently collective, and not just a question of the habits of individuals—the way WE do things. Third, and very importantly, there is a normative element to the expression. It carries the connotation that this is the right, or appropriate, or accepted way to do things. These judgements stem necessarily from shared assumptions, or values, or norms.

The normative element is demonstrated by the reaction of the group to cases of non-compliance. Consider the practice of holding the handrail while descending stairs. If this is indeed the practice in an organisation, there will be a reaction if you fail to do so, ranging from someone reminding you of the rule, to something as unobtrusive as a raised eyebrow. Such reactions may lead a sense of embarrassment or even shame, and can be very effective enforcement mechanisms. Compare this with a situation at my university campus where there are signs saying "cyclists must dismount", but nobody does and there are no consequences. In these circumstances, dismounting cannot be said to be part of the culture, no matter what the university authorities may say. In short, an emphasis on practices does not exclude the importance of norms and values. It just is a question of emphasis.

In my view, then, the most useful way to define culture is as the collective practices of the group—the way we do things around here. The simplicity and concreteness of this expression enables us to avoid most of the conceptual turmoil that surrounds the term. Discussions about culture so often lose their way because culture is an abstract term that rapidly clouds our thinking. As soon as the conceptual fog begins to descend we are less likely to lose our way if we retreat to a more solid reference point: "the way we do things around here".

There is another important reason for preferring this definitional focus when our interest is in changing workplace cultures. Practices can be directly affected by management while values cannot. The organisational anthropologist, Hofstede, puts the point admirably:

> Changing collective values of adult people in an intended direction is extremely difficult, if not impossible. Values do change, but not according to someone's master plan. Collective practices, however, depend on organisational characteristics like structures and systems, and can be influenced in more or less predictable ways by changing these. (quoted in Reason, 1997: 194)

An organisation which focuses its efforts on changing practices is not of course turning its back on value change. Psychology teaches us that human beings feel tension when their behaviour is out of alignment with their values (Kahn, 1984: 115). There is consequently a tendency to bring the two into alignment. If the behaviour is effectively determined by the organisation then the individual's values will tend to shift accordingly. Thus, if an organisation constrains an individual to behave safely, that individual will begin to value safe behaviour more highly. Focussing on practices, therefore, is a not a superficial strategy which leaves the more deep-seated aspects of a culture untouched. Changing practices will in the end change values and assumptions as well. Think, for example, of attitudes to wearing seat belts in cars. When they were first introduced, few people used them. Then they were made compulsory and non-compliers were fined. Accordingly, we changed our behaviour; and over time beliefs themselves changed. Most people now believe it is a good idea to wear seat belts.

Thesis 3 The most useful definition of the culture of a collectivity is its set of collective practices—"the way we do things around here".

4 Description Versus Explanation

Consider the idea of a culture of casual compliance (not *causal* compliance). Such a culture was said to prevail at the BP Texas City Refinery prior to the explosion in 2005 that killed 15 people (Hopkins, 2008: 10). To say that a group has a culture of casual compliance is to make a *descriptive* statement, namely, that people in the group feel no great need to comply with rules and procedures and may do so only when they find it convenient. On the other hand, the statement can be treated as an *explanation* for individual cases of non-compliant behaviour: they occur because of a general culture of casual compliance.

The term "culture of casual compliance" is useful as a *description* because it collects into one category a set of behaviours and attitudes that might not otherwise be linked together. In turn this invites us to explain the phenomenon, using other concepts such as the incentive systems operating in an organisation, or the lack of supervision, or the poor quality of procedures.

On the other hand, treating culture as itself a *cause* of the behaviour of individuals is of limited value, because it offers no insights into the way we might change the culture. It is particularly unhelpful when analysts treat culture as the *root* cause of a problem since this inhibits further inquiry. Moreover if we identify a culture of casual compliance as the root cause of an accident, there is an inevitable tendency to blame the people concerned, which is almost invariably unhelpful, as well as unfair.

Thesis 4 In the organisational context, it is usually better to use culture as a description of group behaviour, rather than as an explanation for individual behaviour.

5 The Sources of Organisational Culture

Having defined organisational culture as the collective practices of the organisation, we can sensibly ask about the source of such a culture. I have at different times given two different answers: structure and leadership.

Consider first the question of structure. The culture of punctuality that exists in many railway systems is an example of how organisational structure creates culture. This culture of on-time-running often requires trains to arrive at and depart from stations within 3 min of the scheduled time. This sometimes results in trains travelling faster than they should in order to maintain schedules; in other words, the culture of on-time running encourages speeding. This was found to be one of the causes of a rail accident causing multiple fatalities near Sydney in 1999 (Hopkins, 2005). The inquiry revealed that this culture was not just a mindset. It consisted of a set of practices which involved people at all levels. Statistics on on-time-running were presented to the senior management twice a day, after each peak hour. Drivers were subject to detailed performance monitoring, and to various sanctions when they failed to meet schedules. There were large numbers of people whose sole job was to ensure that trains ran on time, all of which involved a considerable commitment of resources. It was this organisational apparatus that ensured the pre-eminence of the culture of on-time-running.

The petroleum company, BP, provides a second instructive example. The well blowout in the Gulf of Mexico in 2010 nearly destroyed the company, which determined to change its culture to ensure that this could never happen again. It did so by creating a powerful Safety and Operational Risk (S&OR) function that reported to the CEO. Each geographical business unit had an S&OR manager sitting on its management committee. That S&OR manager was not answerable to the head of that business unit, but to a higher level S&OR manager who answered in turn to someone on the executive committee of the whole BP group. That person reported directly to the CEO of the group. This empowered the S&OR representatives at the local business unit level to stand up to the local business unit leader if they thought it necessary, without jeopardising their careers. The resulting culture gave a greater emphasis to operational excellence than previously. This is a particularly clear example of the way in which "structure builds culture", as an S&OR manager told me, quite unprompted, at interview.

This structural perspective contrasts with a second approach to understanding the source of culture—leadership. Organisational psychologist Edgar Schein puts the point as follows

> Leaders create and change cultures, while managers and administrators live within them. (Schein, 1992: 5)

This is a deliberately provocative statement designed to flatter top leaders into action, but his point is clear enough. If the culture of an organisation is secretive, it is because its leadership has encouraged secretive behaviour; if it is bureaucratic, it is because its leaders have encouraged bureaucratic functioning.

How then do leaders create cultures? I turn again to Schein.

> [Leaders create cultures by] what they systematically pay attention to. This can mean anything from what they notice and comment on to what they measure, control, reward and in others ways systematically deal with.

It is immediately apparent that identifying leaders as the source of culture is not inconsistent with the structural perspective just discussed. The point is that if something is important to top leaders they will set in place the structures that are necessary to ensure the outcomes they want. Leaders create the structures that will in turn institutionalise a certain kind of organisational culture. On-time-running in the rail system is an excellent example of this process.

We must ask finally why it is that top leaders have set in place the structures that in turn create particular cultures. The answer will often lie outside the organisations concerned. For rail systems, the source of concern for on-time-running is public pressure, expressed through various political channels. Sometimes there is even an external regulator that penalises failure to run on time. In the BP case, preventing another major accident became an over-riding concern because of public outrage, as well as the massive financial consequences of the Gulf of Mexico accident. Most importantly, the threat of legal action is a powerful incentive to company officers to put in place structures that will focus attention on safety, and the possibility that CEOs or even directors might be prosecuted has become increasingly real in many jurisdictions.

This external perspective is valuable in counteracting the simplistic view that it all depends on the personal beliefs of the CEO. I have often heard corporate safety managers say that their company is lucky to have a CEO with a passionate personal commitment to safety. Why it is that so many CEOs of global companies today have a passionate commitment to safety, while their counterparts a couple of generations ago apparently had no such commitment? It is hardly likely that the CEOs of today are morally more evolved than those of the past. It is far more plausible that the external environment is now less forgiving of workplace accidents, especially where there are multiple fatalities.

Thesis 5 Organisational cultures depend on the structures that organisations put in place to achieve important outcomes. These structures reflect the priorities of top leaders. The priorities of leaders in turn may depend on factors outside the organisation, such as regulatory pressure and public opinion.

6 Emergent Versus Managerialist Culture

One of the many problematic distinctions in discussions of organisational culture is that between the emergent and managerialist perspectives (Glendon & Stanton, 2000; Haukelid, 2008; Silbey, 2009). These two perspectives are said to have dominated the literature. I touch on this here, ever so briefly, because it has led to so much confusion.

The first perspective, which has its roots in sociology and anthropology, is that the culture of a group is emergent, that is, it emerges from the group in a spontaneous way. On the other hand, the managerialist view, originating in management theory, is that culture is a device that management can use to coerce and control. The first is a bottom up view of culture, while the second is a top down view. These are presented as competing perspectives. The emergent view is sometimes described as an interpretive approach, while the managerialist view is sometimes described as functionalist (Glendon & Stanton, 2000).

This distinction is problematic, however, because it confuses two things: the nature of culture and the origins of culture. We can see this by going back to basics. Culture is the way we do things around here. This presupposes neither an emergent nor a managerialist view. The *origin* of the ways we do things around here is another matter. These ways may well have emerged relatively spontaneously in the group in question, or they may have been engineered by leadership in the manner discussed above. This is surely an empirical question to be determined by investigation. Indeed aspects of the culture may have emerged spontaneously from the group while others have been engineered. If workers at a work site routinely wear hard hats but routinely fail to wear harnesses when working at heights, despite rules requiring them to do so, we can be fairly sure the former practice has been engineered while the latter has emerged from within the group. We don't need to choose at the outset between emergent and managerialist accounts of culture, nor even to adopt some middle position. Rather we can simply ask questions like: what are the limits on leaders' abilities to shape the culture of a work group? The distinction between emergent and managerialist conceptions of culture generates a conceptual fog in which many souls have lost their way.

Thesis 6 The distinction between emergent and managerialist views of culture is misleading.

7 Safety Culture

Finally, safety culture is a term that has led to endless confusion. According to the first and still widely quoted definition of the term, it is an organisational culture in which "safety is an *over-riding* priority" (quoted in Reason, 1997: 194, my emphasis). On the basis of this definition one would have to say that *very few* organisations have a safety culture. As Reason says,

> like a state of grace a safety culture is something that is striven for but rarely attained.

FonCSI, the Foundation responsible for this book, implicitly adopts this position in its very name—Foundation for an Industrial Safety Culture. Here, "safety culture" is being used to describe an aspirational goal, not a characteristic that all organisations have.

On the other hand most users of the term assume that *all* organisations have a safety culture, be it good, bad or indifferent. This is just one of the numerous inconsistencies and confusions that surround the term, since if we accept the definition given above, it makes no sense to speak of a "bad safety culture".

Another source of confusion is that, notwithstanding endless attempt to distinguish between safety culture and safety climate, these two terms are often used interchangeably. (Zohar, 2010, is one writer who uses the safety climate with complete consistency.)

Here is how one recent review summed up the whole situation.

> [Despite all that has been written,] safety culture remains a confusing and ambiguous concept in both the literature and in industry, and there is little evidence of a relationship between safety culture and safety performance. …
>
> Workplace safety may be better served by shifting from a focus on changing 'safety culture' to changing organisational and management practices that have an immediate and direct impact on risk control in the workplace. (SIA, 2014: 8)

This echoes my earlier comments about organisational practices. Notice too that it directs attention to organisational practices without explicitly defining this as the culture of the organisation. In so doing it sensibly sidesteps any definitional debate and goes straight to the heart of the matter.

The question I briefly address here is *why* the term "safety culture" leads to so much confusion. A major reason (there are others) is that the term itself is linguistically problematic. Consider the following compound terms: safety culture, organisational culture, workplace culture, peer-group culture, aviation culture. Safety culture is the odd one out in this list. For all the others, the qualifier—organisational, workplace, etc. —specifies the *group* which is the bearer of the culture. The term says nothing about the *content* of the culture—that remains unspecified. There is thus relatively little scope for confusion. In contrast, with the term "safety culture" the qualifier "safety" does not specify a group. It refers to a quality. (A similar point is made by Schein, n.d.). This is a source of confusion. Does it mean that the culture in question exhibits the quality of safety? If we were to coin the term "punctuality culture" it would have to mean a culture that emphasises punctuality. By analogy, the most natural meaning of safety culture is a culture that emphasises safety. As I have said, this is contrary to the way the term is often used. Safety culture's slide away from its "natural" meaning is facilitated by the fact that safety is a noun, not an adjective. The term "safe culture" would allow no such slippage. It would have to mean a culture that emphasises safety. Clearly, we are now hopelessly entangled in words. And the fault lies not in our thinking; it is the very term "safety culture" that has tied us in knots.

Moreover, this may be a peculiarly English language phenomenon. Neither French nor Spanish have a literal equivalent for "safety culture"; they speak instead of a "culture of safety" (*une culture de sécurité, una cultura de seguridad*), the linguistic implications of which are different. This phrase must surely mean a culture that emphasises safety—a culture that exhibits the quality of safety. If the whole debate about safety culture had occurred exclusively in French or Spanish, I suspect that the primary meaning, indeed the only meaning of *une culture de sécurité* or *una cultura de seguridad* would be a culture that emphasises safety.

I was not dogmatic about safety culture when I first wrote about the concept more than a decade ago. But I did quite deliberately title my book at the time *Safety, Culture and Risk*, not *Safety Culture and Risk.* Today, if I had my way, I would banish "safety culture" from the English language.

Thesis 7 The term safety culture is so confusing it should be abandoned.

Finally, if "safety culture" is abandoned, what terms might be used instead?

If we are talking about a culture in which safety is paramount, then several terms come to mind—a safe culture, a generative culture (Hudson, Parker, & Lawrie, 2006), or even a culture of safety. We can also get away completely from the word culture and talk about mindful organisations (Weick, Sutcliffe, & Obstfeld, 1999), or operational discipline (Angiullo, 2009), or operational excellence (Digeronimo & Koonce, 2016).

On the other hand, if the starting point is that all organisations have a safety culture, then a question like "how good is an organisation's safety culture" can be replaced by "what priority does the organisation give to safety?". Interestingly, in this example, "safety culture" has been effectively replaced by "safety". Or we could ask about risk management practices—a far more down to earth term. Note the word used is practices, not procedures. It is the way we actually do things around here, not the way we are supposed to do things that is of interest.

So all is not lost. There are still plenty of terms available to convey one's intended meaning, whatever it may be.

References

Angiullo, R. (2009). Operational discipline. In A. Hopkins (Ed.), *Learning from high reliability organisations*. Sydney: CCH.

Digeronimo, M., & Koonce, B. (2016). *Extreme operational excellence: Applying the US nuclear submarine culture to your organisation*. Colorado: Outskirts Press.

Glendon, A., & Stanton, N. (2000). Perspectives on safety culture. *Safety Science, 34,* 193–214.

Hale, A. (2000). Culture's confusions. *Safety Science, 34*(1), 1–14.

Haukelid, K. (2008). Theories of (safety) culture revisited—An anthropological approach. *Safety Science, 46,* 413–426.

Hopkins, A. (2000). *Lessons from longford*. Sydney: CCH.

Hopkins, A. (2005). *Safety, culture and risk*. Sydney: CCH.

Hopkins, A. (2008). *Failure to learn: The Texas City refinery disaster*. Sydney: CCH.

Hudson, P., Parker, D., & Lawrie, M. (2006). A framework for understanding the development of organisational safety culture. *Safety Science, 44*(6), 551–562.

Kahn, A. (1984). *Social psychology*. Dubuque: W.C. Brown.

Mearns, K., & Yule, S. (2009). The role of national culture in determining safety performance: Challenges for the global oil and gas industry. *Safety Science, 47*(6), 777–785.

Reason, J. (1997). *Managing the risks of organisational accidents*. Aldershot, UK: Ashgate.

Schein, E. (1992). *Organisational culture and leadership* (2nd ed.). San Francisco: Jossey-Bass.

Schein, E., (n.d.). The culture factor in safety culture. In: G. Grote & J. Carroll (Eds.), *Safety management in context—Cross-industry learning for theory and practice*. Zurich: Swiss Re.

SIA (Safety Institute of Australia). (2014). *The OHS body of knowledge*. Chapter 10, Organisational Culture, Abstract http://www.ohsbok.org.au/wp-content/uploads/2013/12/10.2-Organisational-Culture.pdf?d06074.

Silbey, S. (2009). Taming prometheus: Talk about safety and culture. *Annual Review of Sociology, 35,* 341–369.

Weick, K., Sutcliffe, K., & Obstfeld, D. (1999). Organising for high reliability: Processes of collective mindfulness. *Research in Organisational Behaviour, 21,* 81–123.

Zirngast, E. (2006). *Selective U/W in oil-petro segment*. Unpublished paper.

Zohar, D. (2010). Thirty years of safety climate research: Reflections and future directions. *Accident Analysis and Prevention, 42,* 1517–1522.

Chapter 5
The Safety Culture Construct: Theory and Practice

M. Dominic Cooper

Abstract Safety culture means different things to different people which subsequently guides their improvement efforts. Providing clarity, the essence of the safety culture construct is that it reflects a proactive stance to improving occupational safety and reflects the way people think and/or behave in relation to safety. The extant evidence shows the best proactive stance is to target the significant safety issues found nested within the common safety characteristics (management/supervision, safety systems, risk, work pressure, competence, procedures and rules) identified from public enquiries into process safety disasters. This is best achieved by focusing on the entity's safety management system and their people's safety related behaviours, <u>not</u> by trying to change people's values, beliefs and attitudes. A revised model of safety culture is offered to help guide readers in their quest to improve their safety cultures, along with an adapted model of safety culture maturity. In addition, based on academic evidence and practical experience gained over the past 25 years in numerous industries and countries, the author provides insights into specific issues regarding the influence of senior executives, the impact of national cultures when working on international projects, whether policies and tools should be the same or differ when addressing potential minor, serious and catastrophic events, and who should be involved to drive an organisations safety culture to achieve excellence.

Keywords Safety culture · Safety culture models · Safety culture product
Behavioural factors · Psychological factors · Situational factors
Safety culture maturity

M. D. Cooper (✉)
B-Safe Management Solutions Inc., Franklin, IN, USA
e-mail: info@bsms-inc.com

C. Gilbert et al. (eds.), *Safety Cultures, Safety Models*,
SpringerBriefs in Safety Management,
https://doi.org/10.1007/978-3-319-95129-4_5

1 Safety Culture Theory

The 'safety culture' construct refers to, and is used to, encapsulate and explain organisational safety failings (IAEA, 1991). Its *purpose* is to improve occupational safety in organisations, by preventing low frequency, high severity events such as Chernobyl, Bhopal, Piper Alpha, Texas City, Deepwater Horizon, etc. as well as high frequency, lower impact events (i.e. personal injuries, etc.).

1.1 The Safety Culture Construct

The evolution of any construct proceeds through three overlapping stages (Reichers & Schneider, 1990):

1. *introduction and elaboration* is characterised by attempts to sell the ideas and legitimise the new construct;
2. *evaluation and augmentation* is where critical reviews and early literature on the construct first appear identifying the constructs parameters; and
3. *consolidation and accommodation* is where controversies wane and what is known is stated as a matter of fact.

First introduced in 1984 after the Bhopal disaster, the International Atomic Energy Agency (IAEA, 1991) elaborated on the safety culture construct when defining it as

> *that assembly of characteristics and attitudes in organisations and individuals which establishes that, as an overriding priority, [nuclear power] safety issues receive attention warranted by their significance.*

This clarity led directly to the *evaluation and augmentation* stage. Unfortunately, many influential scholars ignored the IAEA's definition as it did not reflect their 'academic' positions. Under the guise of theoretical purity, academe lost sight of the construct's main purpose—'to stop process safety disasters and serious injuries and fatalities'. There are now more than 50 definitions of the safety culture construct (Vu & De Cieri, 2014) which cause considerable confusion (Hale, 2000) in both industry and academe.

At the heart of these definitional disagreements over the past 30 years or so, is the conflict between interpretive and functionalist approaches. Favoured by social scientists, interpretative approaches state *the organisation is the culture*, where 'cultural' realities are socially constructed solely by the organisations membership. The interpretive emphasis is on gaining an in-depth understanding of the prevailing cultural influences (i.e. assumptions and attitudes) affecting people's behaviour. Conversely, the functionalist approach is favoured by managers and practitioners (the owners of safety culture) who view culture as a variable to be engineered to suit the prevailing circumstances to affect performance by addressing management system faults, people's safety related behaviour, risk-assessments and decision-making.

Almost all of those attempting to define the safety culture construct agree it reflects a proactive stance to improving occupational safety (Lee & Harrison, 2000), and the way people think and/or behave in relation to safety (Cooper, 2000). In reality therefore, most safety culture definitions are functionalist, albeit the interpretive view emphasises shared values, beliefs, attitudes, and norms.

1.2 Influential Safety Culture Models

During the period 1986–2000 three influential models of safety culture were developed to guide theory, research and practice:

1. Guldenmund's (2000) adoption of Schein's (1992) interpretive three-layered organisational culture framework reflecting anthropology and organisational theories;
2. Cooper's (2000) reciprocal safety culture model, based on a functional approach leveraging Social Learning Theory (Bandura, 1977); and
3. Reason's (1998) five inter-dependent sub-cultures (informed, learning, reporting, just, and flexible cultures) based on incident analyses.

Each attempted to provide an actionable framework, and each has been influential in the sense that researchers, regulators and industry have made use of them in some empirical and/or practical capacity.

Guldenmund's (2000) interpretive model contains three layers:

1. *unconscious and unspecified (invisible) core basic assumptions*: the assumptions or suppositions about safety are not articulated, but are taken for granted as the basis for argument or action;
2. *espoused beliefs and values*: operationalised as relatively explicit and conscious 'attitudes' whose targets are hardware (safety controls), software (effectiveness of safety arrangements), people (functional groups) and people's safety-related behaviours; and
3. *artefacts*: visible safety objects (e.g. inspection reports, safety posters, etc.).

In this model, 'culture' is viewed as a pattern of core basic assumptions, invented, discovered, or developed by a group as it learns to cope with external adaptation and internal integration. Explicitly recognising sub-cultures of the overall culture, these differ for executives, engineers and operators. Reducing any significant negative discrepancies between these sub-cultures requires meaningful dialogue between all parties, so they can be explored and minimised by agreeing standardised solutions to practical safety issues (Cooper & Finley, 2013).

Cooper's (2000) functionalist reciprocal model treats safety culture as a sub-culture of an organisations overall culture, while highlighting that it is the *product* of multiple goal-directed interactions between people (psychological), jobs (behavioural), and the organisation (situational). The psychological, behavioural, and situational aspects are the inputs to the safety culture construct, with the key transformation process being the organisations goals, expectations and managerial practices to create the prevailing safety culture *product* (Cooper & Finley, 2013). Formally adopted by the American Petroleum Institute (2015) and the American National Standards Institute (ANSI), the prevailing safety culture is reflected in the dynamic reciprocal relationships between members' perceptions about, and attitudes toward, the operationalisation of organisational safety goals; members' day-to-day goal-directed safety behaviour; and the presence and quality of the organisation's safety systems and sub-systems to support the goal-directed behaviour.

Reason (1998) categorically states safety culture is not a unitary construct as it is made of a number of interacting elements. He equates safety culture with an 'informed culture', which is dependent in turn upon an effective 'reporting culture' underpinned by a 'just culture'. Simultaneously, a 'flexible culture' is required if the organisation is to reconfigure itself in the light of certain kinds of dangers, which in turn will require a 'learning culture'. To some degree these are both objects of, and processes that create, the safety culture *product*: an informed culture.

1.3 Reviewing the Evidence

Cooper's (2016a) recent major review of the safety culture research literature published over the past 30 years showed Guldenmund's interpretive model is clearly *not* linked to actual safety performance. A major conceptual difficulty is that invisible core basic assumptions and/or attitudes are the central core of the safety culture construct, but the evidence shows the link between attitudes and actual safety performance is non-existent to weak. *Thus, in the absence of goals and action, changes in core basic assumptions and attitudes will not stop process safety disasters or serious injuries and fatalities.* Conversely, both Cooper's and Reason's models were clearly linked to actual safety performance. Evidence reveals companies should focus at least 80% of their culture change efforts on situational (e.g. safety management systems) and behavioural factors to prevent process safety disasters and Serious Injury and Fatality (SIF) incidents. For example, top management can positively influence an organisations safety culture by paying attention to the *effectiveness* of the development, execution, and performance of the safety management system, frequently questioning managers about safety matters during routine meetings, and frequently visiting various business units to discuss safety issues.

Attempting to *consolidate and accommodate (stage 3) to make clear what is known*, making use of the IAEA's (1991) definition of safety culture as a framework, Cooper's (2016a) explored its constituent parts. What are the assembly of safety culture characteristics? The assembly of attitudes? The significant safety issues?

Consensus was found on six major safety culture characteristics when examining academic research and the results of public enquiries into process safety disasters:

1. management/supervision;
2. safety systems;
3. risk;
4. work pressure;
5. competence; and
6. procedures and rules.

Typically, each of these characteristics are contained in modern safety management systems (e.g. OSHA (S) 18001:2007; ANSI-Z10: 2012) implemented in many countries. Already aligned with existing practical and proven Health, Safety and Environment (HSE) strategies and processes, companies should prioritise these safety culture characteristics to effect change.

Respectively reflecting the interpretive and functional perspectives of safety culture, academe tends to emphasise the associated psychological factors of these six characteristics, whereas the results of public enquiries focus almost entirely on improving tangible situational and behavioural factors within company operations. Although there were at least twelve psychological factors to target to influence safety culture change, none were found to be consistently and reliably linked to actual safety behaviour or actual safety incident rates. Instead, results revealed a sole focus on psychological factors when changing and/or assessing safety culture (a common approach in industry) is fatally flawed. Conversely, both situational and behavioural factors were clearly linked to actual safety performance. As such, companies are urged to stop focusing on psychological factors and concentrate their improvement efforts on tangible situational and behavioural factors.

Common significant safety issues within each of the six safety culture characteristics were also explored by examining a series of independent studies into the causal factors across numerous process safety incidents (e.g. Collins & Keely, 2003). This showed 80% of Loss of Primary Containment incidents (LOPC's) are commonly caused by managerial behaviours, or lack of, and that 80% of process safety disasters occur during normal routine everyday operations (64%) and maintenance (16%). Depressingly, similar managerial behaviours were also found to be related to the occurrence of SIFs. Such dramatically clear findings show safety leadership has to become a fundamental managerial competency.

Figure 1 presents the universally applicable targets of safety culture (i.e. its characteristics and the significant safety issues associated with each) identified by Cooper incorporated into a revised reciprocal safety culture model (Cooper, 2016a). It shows companies should focus on the common root causes of Process Safety and SIF incidents to drive desired behaviour. *The principle is to optimise the situation to optimise the behaviour*. In turn, as the desired behaviours become habitual, the various psychological factors will become more positive.

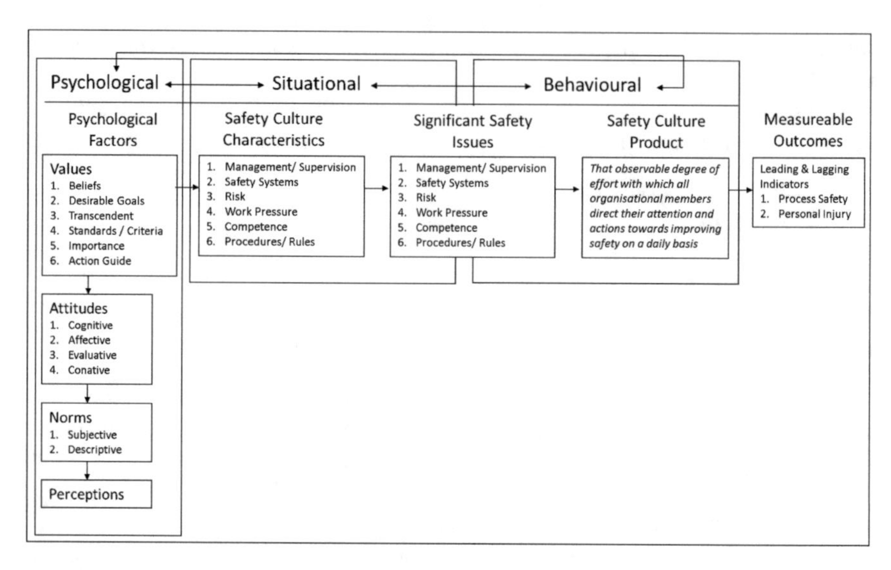

Fig. 1 Cooper's (2016a) revised reciprocal model of safety culture

Summarised below, a focus on the salient issues and the evidence-based solutions to address them that would change company's safety cultures for the better (see Cooper, 2016a for details) are:

The '*Management and Supervision*' characteristic is primarily concerned with people's visible safety leadership: ineffective safety leadership often stems from confusion about (a) the company safety management systems and associated policies; (b) a leader's individual safety responsibilities and obligations; (c) the leader's and others' authority over safety; and (d) what leaders are being held accountable for (Cooper & Finley, 2013). This confusion has often led to managers and supervisors failing to adequately plan activities, not managing the safety of operations, and not being adequately prepared for an incident (e.g. IAEA, 2014). It would help if companies developed Leadership Behavioural Competency and Accountability Matrices defining its managerial and supervisory roles and responsibilities, clarifying what people are expected to do and when, with associated performance measures being used to ensure leaders are doing the right things at the right time, for the right reasons.

The '*Safety Systems*' characteristic refers to any formalised *strategic* system to control HSE. Based on the LOPC research, however, it is argued the primary areas of opportunity consist of optimising: (a) two-way safety communications processes; (b) incident analyses and lessons learned processes; (c) the design of plant, equipment, and processes so that safety is an integral element; (d) asset integrity to ensure material conditions meet the expected standards; and (e) management of change processes to ensure they are related to risk assessment and analysis. Each of these require clear policies and procedures.

The '*Risk*' characteristic refers to (a) risk appraisal; (b) risk assessment; and (c) risk controls. The LOPC research points to a significant number of failures in each of these areas, indicating that the 'Risk' characteristic represents a fundamental weakness in the majority of companies.

The '*Work Pressure*' characteristic primarily refers to the safety-production conflict that stems from competing priorities, lack of resources or of a willingness to treat *safe production as the number one priority*. The costs of incidents tend to outweigh any perceived advantages of placing productivity before safety (HSE, 2016), but this is often overlooked by managers trying to satisfy their immediate job-related needs. This is one area where a company's top management team can unequivocally stamp its authority on its managers and operators, by setting the right expectations and reinforcing them through an alignment of their Key Performance Indicators (KPIs) and productivity bonus systems.

The '*Competence*' characteristic refers to the knowledge, skills, and abilities people possess to do their job efficiently and effectively. From a process safety perspective, poor competence is often revealed in control rooms when operators fail to recognise and react to early warning signals and/or adequately respond to incidents. Similarly, plant personnel often misuse or incorrectly operate equipment and/or fail to complete isolations properly. It is imperative that people are sufficiently trained in the safety aspects of their jobs to the point that they cannot get things wrong. Currently, people often receive training only until they get something right. In essence, rehearsal is the key to developing people's competence.

The '*Procedures/Rules*' characteristic refers to all those codified behavioural guidelines developed by companies to form their safety management system. In too many cases, process safety catastrophes and SIFs stem from (a) an absence of procedures (e.g. a lack of procedures altogether, or those developed are not freely available to the workforce); (b) the presence of poor quality procedures; and (c) a lack of procedural reviews. These situations lead to non-compliance, where managers tend to circumvent the administrative aspects of safety, or put productivity before safety, while employees tend to circumvent them to make their task easier in some way. Clearly, the way forward is to (a) identify any gaps in written procedures; (b) allow the workforce to review the existing procedures to ensure they are safe, they make sense and are easily understood; (c) monitor procedural compliance, and (d) regularly audit those procedures involved in near-miss incidents or accidents.

Incorporated into the model presented in Fig. 1, the safety culture product,

> *that observable degree of effort with which all organisational members direct their attention and actions towards improving safety on a daily basis* (Cooper, 2000)

provides a universal measure of safety culture with which to assess the impact of change (i.e. are people putting in more effort to improve safety as a result of an intervention?). Evidence (e.g. Vogus & Sutcliffe, 2007) shows this *product* is a viable and practical means of measuring safety culture. The results can be graded

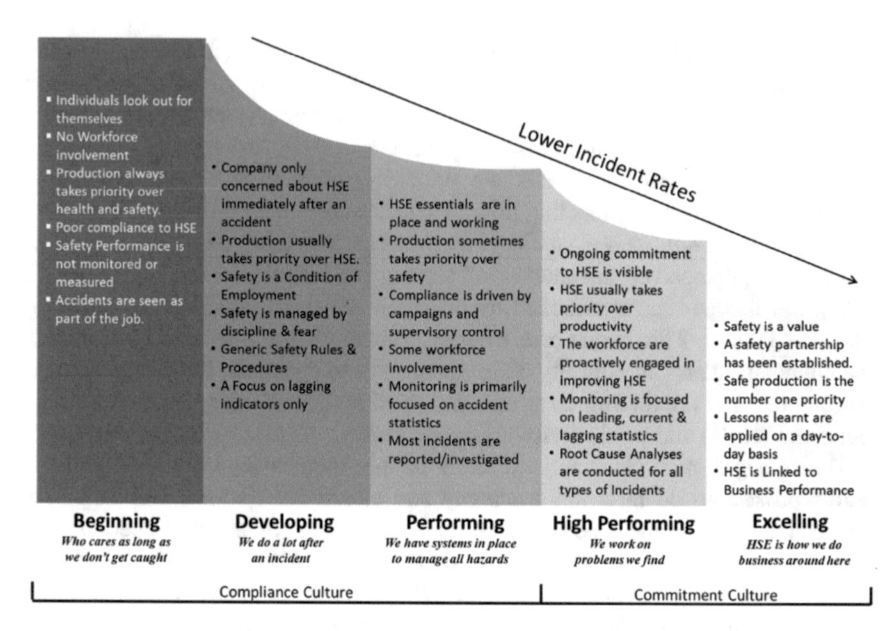

Fig. 2 Adaptation of the British health and safety executive's (2011) safety culture maturity model

against safety culture maturity models (e.g. HSE, 2011) which facilitate benchmark assessments within a company or against others in industry. Typically, these are divided into five safety culture maturity levels (see Fig. 2) specifying an organisations level of effort (e.g. Beginning, Developing, Performing, High Performing, and Excelling) as it progresses on its safety culture improvement journey, and are, therefore, *de facto* measures of the safety culture product, "*that observable degree of effort…*".

Safety culture assessments are typically conducted on an annual or bi-annual basis. In the interim, it makes good commercial sense to develop leading KPIs that focus on the safety characteristics outlined above, but *with an emphasis on what people do*, so that the level of effort put into safety (i.e. the safety culture product) can be easily monitored. For example, KPIs for the 'Management and Supervision' characteristic could include (a) the number of corrective actions completed with 30 days; (b) the number of safety observations and conversations a manager/supervisor had with the workforce each week. Such measures facilitate and enable transparency. In turn, this allows companies to monitor the integrity, and maintenance of its improvement initiatives, while also revealing the status of their safety culture *product.*

2 Safety Culture Practice

Given the *purpose* of the safety culture construct is preventing process safety and personal injury incidents, what should industry do to help ensure this? Typical questions are:

To what extent can changes in safety culture be achieved as a result of decisions by top management?

Executive level managers and board members have to prioritise and balance safety against production, stock-market concerns and other commercial/operational/political pressures. The expectations they set, the management practices they reinforce, and the performance outcomes they reward (i.e. shaping the situational aspects) will all influence the safety culture. ExxonMobil provides a true example of ensuring safety is an integral part of their operating culture, where they strongly believe protecting the safety and health of their workforce is fundamental to its business. In 2007, ExxonMobil drillers in the Gulf of Mexico asked if they could stop drilling the Blackbeard West at 30,000 ft (the goal was 32,000) as they felt it was too dangerous to drill deeper, having experienced a 'kick' that made the platform tremble. The prize was over a billion barrels of oil and the geologists wanted to continue because of the rewards on offer. The decision was pushed right up to the CEO, who erred on the side of safety, saying the 'Well' had only cost $180 million dollars to date. He received strong criticism from Wall Street but no lives were lost. In contrast, the Deepwater Horizon drillers were ignored when they expressed similar concerns about the Macondo Well. This ultimately cost 11 lives, the loss of a platform, an environmental disaster, with BP's costs and fines reaching $42 Billion to date. Clearly, executive level managers who consider the safety element in all their decisions can guide and impact others decision-making and actions to prevent disasters and personal injuries. This again points to the fundamental importance of safety leadership in everyday operations: ensuring safety before profit, cultivating a trusting and fair culture, making decisions that err on the side of safety, developing safety competencies for all, applying lessons learned, ensuring compliance to well-written rules and procedures, and constantly communicating meaningful safety messages. If senior managers do not manage and reinforce these issues, their company's safety culture will never achieve excellence.

What are the relative influences of the national/local culture, corporate culture, and professional cultures, on the safety culture of a given entity?

A study in the global Oil and Gas industry showed western countries tend to have higher risk tolerance and higher incident rates than Asian countries, who have much higher respect for authority (Brown, 2012). Total Recordable Incident Rates were much lower on projects where the site safety culture embodied a combined 'high perception of risk' and 'low tolerance of risk', compared to those with a high/low perception of risk and high tolerance for risk. Other work (e.g. Perez-Floriano & Gonzalez, 2007) shows there has to be respect for national culture traits by working with them if risk management programs are to be successful. National cultural differences reside mostly in values, while at the organisational level, cultural differences reside mostly in practices (Hofstede, 1983) suggesting that national cultures can be over-ridden by the company's practices and their prevailing safety culture. This is reinforced by Mearns and Yule (2009) who found that proximal influences such as perceived management commitment to safety and the efficacy of safety measures exert more impact on workforce behaviour and subsequent accident rates than fundamental national values. Another example showed exemplary safety leadership practices in conjunction with genuine employee engagement on a middle-east construction project with 47,000 third-party nationals from 64 countries led to 121 million man-hours worked without a single lost-time accident (Cooper, 2010). Thus, the key ingredient for success is the quality of safety leadership at the local level to ensure risk management initiatives are implemented effectively. However, expats who provide local site leadership must be provided with the tools and skills needed to address a broad spectrum of local cultural needs.

Should the safety culture be the same in the whole corporate organisation, or should it be implemented diversely according to local activities/cultural features?

There have always been sub-safety cultures (even in the same facility) which is likely due to each group's differing 'frame of reference' for viewing the risks presented by tasks (Cooper, 1997). The real issue is whether this should be catered for in some specific way. In the author's experience, the role of the corporate executive team is to provide a clear framework for action that sets the parameters, but allows for some degree of local variation: the main point being that people are doing things to improve safety within the parameters set. This approach, commonly known as '*pull and push*', is where a global framework (i.e. policy, template tools, roles and accountabilities, resources, time frames, etc.,) is provided by the corporate offices, but these are tailored and implemented to suit local conditions. The underlying principle, therefore, is to do safety *with* people, not *at* them. This is also where the participation of engaged employees comes into its own, as they are intimately familiar with all aspects of their work and can provide insights often overlooked by corporate safety departments and managers.

Should occupational safety regarding minor risks, the prevention of fatalities and the prevention of major industrial risks be managed with the same policy and the same tools?

In principle, the policies that govern occupational safety to prevent and protect people from workplace hazards and risks are broadly similar: they attempt to define a problem area, assess its scope, and give direction on the control of the issues. However, the tools and strategies required to control the various aspects of Safety and Health will differ. Recent work on SIFs shows that the causes of life-threatening and life-altering events tend to be different than those for minor personal injuries (Cooper, 2014; Wachter & Ferguson, 2013). Thus, a specific SIF program targeting *potential* SIFs is required. The same can be said for process safety with its emphasis on the blending of engineering and management approaches, as event outcomes are very different from personal injuries, albeit the underlying managerial causes tend to be the same for both. A good/poor safety culture affects all managerial aspects of Occupational Safety and Health.

Who should drive an organisation's safety culture to help it evolve?

Two key initiatives (i.e. situational changes) are known to drive an organisation's safety culture to achieve safety excellence: safety leadership and employee engagement, within a formal ethos of developing a '*safety partnership*'. Both are contained within the 'Management/Supervision' characteristic in the model shown in Fig. 1, and lend themselves to monitoring the safety culture product, "*that observable degree of effort...*".

2.1 Safety Leadership

Defined as

> *The process of defining the desired state, setting up the team to succeed, and engaging in the discretionary efforts that drive the safety value* (Cooper, 2015),

safety leadership is widely recognised to be extremely important, especially when the prevailing safety culture is weak (Martínez-Córcoles, Gracia, Tomas, & Peiro, 2011). A company's safety culture is driven by the executive leadership team who creates, cultivates, and sustains its journey to excellence. They set the vision and the strategic direction (i.e. *the desired state*), provide resources (i.e. *set up the team to succeed*), and constantly emphasise and reinforce the importance of safety to people and the business (i.e. *engage in the discretionary efforts to drive the safety value*). For a variety of reasons, ineffective safety leadership is a major blockage to achieving success in many companies (Cooper & Finley, 2013).

Recent research, summarising 328 safety leadership studies, examined the impact of transformational, transactional and servant leadership styles on actual safety performance (Cooper, 2015). All three styles directly influence people's safety behaviour, which in turn reduce incident rates. However, the positive effects were stronger for servant leadership. The major difference is a servant leadership style naturally creates a supportive environment that exerts strong direct influences on employee engagement, safety behaviour, and incident reduction, which the transformational and transactional safety leadership styles do not. In practice, leaders who engage in meaningful two-way dialogues with the workforce induce a collaborative learning environment and facilitate other people's safety needs, helping to create the supportive environment that appears so important for improving safety performance.

Unfortunately, there are always two sides to a coin: the more hazards and risks that are present in the working environment, the lower the impact of any safety leadership style and the bigger the barriers to creating a supportive environment. If a company's safety leadership efforts are to flourish, it is imperative that a supportive environment is also developed for managers, and sufficient resources provided so they can eliminate or reduce known hazards/risks to as low as reasonably practicable (ALARP).

2.2 Employee Engagement

The business benefits to be obtained from employee engagement are huge. Studies have shown that (a) where employee engagement was low, companies had 62% more safety incidents (Harter, Schmidt, Killham, & Asplund, 2006); and (b) where employee engagement was high, engaged employees were five times less likely to experience a safety incident, and seven times less likely to have a lost-time safety incident (Lockwood, 2007) than non-engaged employees. Employee engagement is an approach designed to help ensure employees are committed to an entity's goals and values, while motivating people to contribute to that entity's success. Such entities tend to possess a strong and genuine value for workforce involvement, with clear evidence of a 'just and fair' culture (Reason, 1997) based on mutual respect between the entire management structure and the workforce. The key aspect is ensuring an understanding by all concerned that engagement means two-way dialogues that lead to joint decision-making about the best way forward, while also acting together to make things happen: managers deliberately reach out to engage with employees to focus on issues of importance (e.g. safety), who in turn proactively and positively engage with management. In sum this means creating a genuine *safety partnership* between management and the workforce to improve safety performance.

2.3 A Safety Partnership

A safety partnership is defined as:

> *Leadership, managers and front-line associates jointly focusing on safety and proactively working together in a business entity to minimise the possibility of harm and maximise safety performance.* (Cooper, 2016b)

Creating a genuine safety partnership, therefore, means management and the workforce jointly working towards achieving common and understood safety goals, with clear and consistent communication, efficient monitoring, reporting, and decisive action to investigate blockages and take the appropriate corrective action as needed.

The key drivers for developing and maintaining a safety partnership are straightforward and involve (a) effective safety leaders who develop a supportive environment; and (b) reducing the degree of risk presented by the nature of the work. High levels of managerial support lead to higher levels of engagement, which in turn lead to much higher compliance with safety rules and procedures. Moreover, reducing levels of risk presented by hazards and high job-pressures also leads to much higher compliance with safety.

Specific areas of safety that joint management and workforce teams can use to develop a proactive safety partnership include: (a) safety leadership skills development; (b) hazard identification exercises; (c) risk assessments; (d) reporting, investigating and reviewing incidents; (e) reviews of rules and procedures; (f) employee development of toolbox talks; (g) mentoring new hires; (h) pro-active involvement in behaviour-based safety processes; and (i) seeking people's views on improving safety.

3 Summary

'Safety culture' is a social construct used by industry and academe to describe the way that safety is being managed in organisations to avoid catastrophes and personal injuries. As well as being used to save lives and prevent process safety disasters, it is known that operational and safety excellence go hand-in-hand; companies that are good at managing safety also manage operations well (Fernández-Muñiz, Montes-Peón, & Vázquez-Ordás, 2009; Veltri, Pagel, Behm, & Das, 2007).

In terms of safety culture theory, almost all of those attempting to define the safety culture construct agree that it reflects a proactive stance to improving occupational safety, and the way people think and/or behave in relation to safety. As such, these should be treated by industry as the key underlying factors that guide their improvement efforts.

A major review of the evidence (Cooper, 2016a) showed: (a) there is consensus between academe and the results of public enquiries about the main safety culture characteristics a company should target to improve its organisational safety culture; (b) the sole use of psychological safety surveys to assess a company's safety culture is fatally flawed as they are not reliably linked to actual safety performance; (c) common significant safety issues to avoid process safety disasters and SIFs are well known, and provide a tangible and robust focus for assessing the safety culture construct; (d) organisations should concentrate 80% or more of their safety culture improvement efforts on situational and behavioural (e.g. managerial safety related leadership behaviours) factors to prevent process safety and SIF incidents; and (e) the safety culture product should be used to assess safety cultures, the results of which can be used to determine a company's safety culture maturity. Companies should develop leading KPIs that focus on what people do, to facilitate the monitoring of "*that observable degree of effort…*".

In terms of safety culture practice, evidence shows that: (a) senior executives have to consider the safety element in all their decisions to guide and impact other's decision-making and actions to prevent disasters and personal injuries; (b) the quality of safety leadership at the local level to ensure risk management initiatives are implemented effectively tends to override national culture considerations; (c) every organisation *will* have sub-safety cultures, and adopting a '*pull and push*' approach where a corporate framework is provided that can be tailored and implemented to suit local conditions, is the best way forward; (d) different policies and tools are needed to address minor, major, and catastrophic events; and

(e) creating a safety partnership that fully involves both management and employees in the safety improvement effort is the best way for an organisation's safety culture to evolve and achieve excellence.

References

American Petroleum Institute. (2015). *Pipeline safety management systems standard* (ANSI/API RP 1173).

Bandura, A. (1977). *Social learning theory*. Englewood Cliffs, NJ: Prentice-Hall.

Brown, C. (2012). *Kentz Group Safety Conference*. Shaping future safety culture: Learning from shared best practice. Retrieved from: http://www.kentz.com/media/101231/kentz___safety_conference_cards_2012.pdf.

Collins, A., & Keely, D. (2003). *Loss of containment incident analysis*. HSL/2003/07.

Cooper, M. D. (1997). Evidence from safety culture that risk perception is culturally determined. *The International Journal of Project & Business Risk Management, 1*(2), 185–202.

Cooper, M. D. (2000). Towards a model of safety culture. *Safety Science, 36,* 111–136.

Cooper, M. D. (2010). Safety leadership in construction: A case study. *Italian Journal of Occupational Medicine and Ergonomics: Supplement A: Psychology, 32*(1), A18–A23.

Cooper, M. D. (2014). Identifying, controlling and eliminating serious injury and fatalities. In H. Beach (Ed.), *Beyond compliance: Innovative leadership in health and safety* (pp. 23–29). SHP/UBM.

Cooper, M. D. (2015). Effective safety leadership: Understanding types & styles that improve safety performance. *Professional Safety, 60*(2), 49–53.

Cooper, M. D. (2016a). *Navigating the safety culture construct: A review of the evidence*. Franklin, IN, USA: BSMS.

Cooper, M. D. (2016b). Practical employee engagement. In *ASSE Safety 2016 Professional Development Conference & Exposition*, June 26–29, Atlanta, GA.

Cooper, M. D., & Finley, L. J. (2013). *Strategic safety culture roadmap*. Franklin, IN, USA: BSMS.

Fernández-Muñiz, B., Montes-Peón, J. M., & Vázquez-Ordás, C. J. (2009). Relation between occupational safety management and firm performance. *Safety Science, 47*(7), 980–991.

Guldenmund, F. W. (2000). The nature of safety culture: A review of theory and research. *Safety Science, 34,* 215–257.

Hale, A. (2000). Culture's confusions. *Safety Science, 34,* 1–14.

Harter, J. K., Schmidt, F. L., Killham, E. A., & Asplund, J. W. (2006). *Q12® Meta-analysis*. Washington, DC: Gallup Consulting.

Health and Safety Executive. (2011). *Development of the people first toolkit for construction small and medium sized enterprises*. RR895. HSE Books.

Health and Safety Executive. (2016). *Costs to Britain of workplace fatalities and self-reported injuries and ill health, 2014/15*. HSE Books.

Hofstede, G. (1983). The cultural relativity of organisational practices and theories. *Journal of International Business Studies, 14*(2), 75–89.

IAEA. (2014). *Nuclear safety review 2014*. GC (58)/INF/3. Vienna: IEAE.

International Atomic Energy Agency. (1991). *Safety culture. A report by the international nuclear safety advisory group*. Safety series (75-INSAG-4). Vienna, Austria: IAEA.

Lee, T., & Harrison, K. (2000). Assessing safety culture in nuclear power stations. *Safety Science, 30,* 61–97.

Lockwood, N. R. (2007). Leveraging employee engagement for competitive advantage: HR's strategic role. *HR Magazine, 52*(3), 1–11.

Martínez-Córcoles, M., Gracia, F., Tomas, I., & Peiro, J. M. (2011). Leadership and employees' perceived safety behaviors in a nuclear power plant: A structural equation model. *Safety Science, 49,* 1118–1129.

Mearns, K., & Yule, S. (2009). The role of national culture in determining safety performance: Challenges for the global oil and gas industry. *Safety Science, 47*(6), 777–785.

Perez-Floriano, L. R., & Gonzalez, J. A. (2007). Risk, safety and culture in Brazil and Argentina: The case of TransInc Corporation. *International Journal of Manpower, 28*(5), 403–417.

Reason, J. (1997). *Managing the risks of organizational accidents.* Aldershot, Hants: Ashgate Publishing.

Reason, J. (1998). Achieving a safe culture: theory and practice. *Work & Stress, 12*(3), 293–306.

Reichers, A. E., & Schneider, B. (1990). Climate and culture: an evolution of constructs. In B. Schneider (Ed.), *Organisational climate and culture.* San Francisco, CA: Jossey-Bass.

Schein, E. H. (1992). *Organisational culture and leadership* (2nd ed.). San Francisco, CA: Jossey-Bass.

Veltri, A., Pagel, M., Behm, M., & Das, A. (2007). A data-based evaluation of the relationship between occupational safety and operating performance. *The Journal of SHE Research, 4*(1), 1–22.

Vogus, T. J., & Sutcliffe, K. M. (2007). The safety organizing scale: Development and validation of a behavioral measure of safety culture in hospital nursing units. *Medical Care, 45*(1), 46–54.

Vu, T., & De Cieri, H. (2014). *Safety culture and safety climate definitions suitable for a regulator: A systematic literature review* (Research report 0414-060-R2C). Monash University.

Wachter, J. K., & Ferguson, L. H. (2013). Fatality prevention: Findings from the 2012 forum. *Professional Safety, 58*(7), 41–49.

Chapter 6
A Pluralist Approach to Safety Culture

Safety Cultures as Management Tools and as Professional Practices

Benoît Journé

Abstract Managing safety culture appears to be a very difficult task, including in the context of high-risk industries. A clear opposition exits between academics about this issue. On the one hand some deny the possibility for an organization to "manage" any kind of culture. Doing so would just be a manipulation of groups' and individuals' behaviors that has nothing to do with culture but refers to coercive power and domination. On the other hand, some build up theoretical frameworks and good practices to support the development and the maintenance of a strong and homogenous organizational culture such as safety culture. Our contribution to this debate is to open a way between these two opposite approaches. The aim is to introduce a pluralist approach of safety culture that makes its management possible, meaningful and valuable for both managers and practitioners. It is based on the clear distinction between two sets of safety cultures: Safety-Culture-as-Tools (SCT) and Professional-Safety-Cultures (PSCs).

Keywords Safety culture · HRO · Integration · Differentiation
Professional and occupational communities · Tools · Boundary objects
Discussion spaces

1 Two Types of Cultures: Safety-Culture-as-Tools (SCT) and Professional-Safety-Cultures (PSCs)

Safety-Culture-as-Tools (SCT) is a set of management tools designed to create a single "organizational" safety culture. The formal Safety Culture as promoted by the IAEA in the nuclear industry (INSAG 4) in 1991 and officially adopted and implemented since then can actually be defined as a "management tool". This culture

B. Journé (✉)
IAE Nantes, Université de Nantes, Nantes, France
e-mail: benoit.journe@univ-nantes.fr

B. Journé
IMT Atlantique, Chaire RESOH, Nantes, France

C. Gilbert et al. (eds.), *Safety Cultures, Safety Models*,
SpringerBriefs in Safety Management,
https://doi.org/10.1007/978-3-319-95129-4_6

is supposed to be learned, shared and implemented by every individual in the organization. It is an espoused set of homogeneous values and formal practices oriented towards safety and mostly defined and enacted by the top and middle-managers of the organization. It is embedded in various techniques and tools, such as "risk assessment", "questioning attitude", "transparency" … but also in official discourses about safety, such as "safety first". This Safety-Culture-as-Tools is also represented by the formal safety indicators and the ways they are used to balance other performance indicators.

Considering safety culture as a management tool supposes to have a clear representation of what a management tool is. A management tool is an artifact that promotes, influences and controls the actors' behaviors in order to achieve a certain goal, which has been set by the managers or by the organization they belong to. Formal procedures of risk analysis are good examples of such SCT.

One of the main advantages of SCT lies in its rationality, homogeneity and alignment with the strategic and managerial orientations of the organization. Moreover, this kind of safety culture, mostly based on written documents, can easily be assessed and audited.

This also has limits and drawbacks, the main one being the possible lack of legitimacy and relevance. Practitioners who acquired safety expertise through their day-to-day activities may consider the official and formal safety culture as inadequate, because it is much too far from the realities on the ground. For them, SCT may just become a bureaucratic burden that makes it difficult to do a good job. A gap can progressively appear between this "espoused" SCT and the safety culture "in use" in practitioners' communities.

By contrast, Professional-Safety-Cultures (PSCs) are multiple and located in working groups and professional and occupational communities. They encompass the knowledge, values, attitudes and practices created and mobilized in order to "do a good job" in a risky environment. They emerge through time, from shared experiences and evolve with collective learning processes. PSC is the expression of the ability of a group to successfully mix safety with other dimensions of industrial performances (faster, better, cheaper…) in their daily decisions and practices. For many academics it is the only genuine form of "safety culture".

The multiplicity of PSC echoes the multiplicity of teams and communities of practice present in complex organizations. It is a key resource for the reliability and resilience of HRO[1] (Weick, 1987). Indeed, based on a "sensemaking" perspective, the complexity and variety of the unexpected problems call for a high level of decentralization, diversity and differentiation within the socio-technical system just to make sure that people on the ground understand (make sense of) what is occurring, make the right decision and take the right action as quickly as possible. This is the main advantage associated with PSCs (Antonsen, 2009).

[1]High Reliability Organization.

Table 1 Safety-Culture-as-Tools and Professional-Safety-Cultures

	SCT	PSCs
Origins and legitimacy	External knowledge, expertise and principles	Professional groups, day-to-day activities
Forms	Formal guides, tools and practices	Practices and expertise of professionals, technicians and practitioners
Organizational alignment	Guaranteed by the top-down approach	Not guaranteed due to the bottom-up approach
Associated risks	Lack of relevance = creation of a "fake" safety culture	Lack of coherence

The limits of PSCs lie in their heterogeneity which may lead to potential horizontal conflicts between professional groups, and vertical conflicts with managers because they offer no managerial alignment. PSC are socially regulated but not easily controllable and manageable (Table 1).

2 The Complex Relationships Between SCT and PSCs

The existing literature opposes these two safety cultures, with SCT and PSCs competing for legitimacy and dominant position in the organization. SCT is often seen as a negation of PSCs, either as a result of managerial ignorance or by the will to tighten control over the employees. Therefore, the development of a strong formal SCT often creates tensions between managers and professionals or occupational groups. The latter tend to resist, appear reluctant to adopt the official SCT and defend their own PSCs. Managers tend to interpret it as a kind of "resistance to change", a lack of rigor and safety knowledge that requires more training, management control and command. A vicious circle of mutual misunderstanding is at play. The transaction between SCT and PSCs is blocked and so is the possibility of improvements. We suggest that this vicious circle can be transformed into a positive interaction between SCT and PSCs, especially in the case of HRO.

The outcome of the competition between SCT and PSCs depend on their relative weights. Table 2 presents four configurations based on the combination of weak/strong SCT and weak/strong PSCs.

The first configuration is characterized by a lack of Safety Culture due to a weak SCT combined with weak PSC. It includes neither global nor local management of safety issues. It may exist in many industries but is unacceptable in high risk industries.

The second configuration is characterized by the domination of PSCs over SCT. The richness and diversity of the PSCs help the organization tackle safety issues within their local boundaries. But organizational problems arise when several professional safety cultures compete because several communities of practices

Table 2 Four configurations of safety cultures

SCT	PSCs	
	Weak	Strong
Weak	(1) Lack of safety culture Low level of differentiation Low level of integration Vulnerability: unacceptable in high-risk industries	(2) Professional safety cultures High level of differentiation Low level of integration Vulnerability: lack of coherence, multiplication of conflicts, no strategic alignment
Strong	(3) Bureaucratic safety culture High level of integration Low level of differentiation Vulnerability: lack of relevance, inability to cope with complex problems	(4) HRO safety culture Highly differentiated and highly integrated Vulnerability: requires important organizational slack that may be threatened by rationalization programs (cost cutting…)

disagree about the diagnosis or solutions for the issue at stake. Without a strong integration process, this situation may lead to horizontal conflicts between professionals and the impossibility to build up acceptable compromise.

The third configuration is a bureaucratic Safety Culture generated by a domination of SCT over PSC.

The fourth configuration is the HRO Safety Culture characterized by the coexistence of a strong SCT and strong PSCs. As seen before, the competition between SCT and PSC can end up with a "vicious circle" of mutual delegitimization. The HRO model opens a way for a balance between a strong SCT and strong and multiple PSCs. Even if SCT and PSCs are potentially conflicting, we suggest that organizations such as HRO that require a strong and genuine safety culture are actively managing the combination of a strong SCT and a strong SCP.

Our analysis echoes the four types of safety cultures identified by Daniellou, Simard, and Boissières (2010, p. 102). Considering the importance of employee and management commitment for safety, they distinguish the "fatalist culture" (low level of commitment of both employees and management), the "integrated culture" (high level of commitment of both employees and management), the "management culture" (low commitment of employees but high for management) and the "professional culture" (high commitment of employees but low for management). Nevertheless, our analysis appears to be less focused on the level of commitment for safety than on the level of differentiation and integration of the safety cultures and the interactions and dialog between differentiated professional safety cultures and the integrated SCT.

3 Organizing the Dialog Between PSCs and SCT

Following the Lawrence and Lorsch (1967) differentiation/integration model, we suggest that strong PSCs require strong SCT. In this model, the most successful firms competing in complex environments are both—and simultaneously—highly "differentiated" and highly "integrated". Lawrence and Lorsch define differentiation as the

> state of segmentation of the organizational systems into subsystems, each of which tends to develop particular attributes in relation to the requirements posed by its relevant external environment.

A high level of differentiation is the organizational solution for remaining efficient in complex and changing environments. In a differentiated organization, each part develops its own skills, knowledge, ways to do things and finally its own language and culture. When the system becomes highly differentiated, the organization faces a substantial risk: the progressive lack of internal coherence, due to the growing fragmentation of the subsystems and the multiplication of internal misunderstandings and conflicts which may lead to a loss of control over the organization and even to a potential breakdown. Thus, integration is required to prevent this risk. Integration is defined as

> the process of achieving unity of effort among the various subsystems in the accomplishment of the organization's task.

Management control systems and reporting practices are classical integration processes. Lawrence and Lorsch also mentioned the "organizational culture" as an important integration factor.

Applied to the sphere of safety culture, the various PSCs play the role of differentiation whereas SCT plays the role of integration. The stronger the PSCs become, the higher the risk of unsolved conflicts between professional groups is expanding. This calls for a strong integration mechanism. SCT may play this role. SCT can be a way for the professional groups to solve the conflicts they may have about safety valuations, diagnosis or solutions. Indeed, the dialog between PSCs is neither spontaneous nor easy to achieve when disagreements appear about safety issues.

The dialog has to be organized in order to build up acceptable local and temporary compromises between various competing communities of practice. In this perspective, two methods can fruitfully be explored. The first one is the design and implementation of "discussion spaces" (Detchessahar, 2013; Rocha, Mollo, & Daniellou, 2015) allowing the practitioners and the managers to discuss safety issues from their own professional safety cultures. The second way that may be interesting to explore is to see SCT as a "boundary object" (Star & Griesmer, 1989). Such objects (like maps, procedures…) facilitate the coordination and dialog between various professional groups and communities of practice (Tillement, Cholez, & Reverdy, 2009). They help them to deal with the internal boundaries of the organization. The SCT (the different tools in which the SCT is embedded) may play the role of a boundary object for the different PSCs. In a sense, SCT is a kind of common language, focused on the minimal safety assumptions promoted by the management and shared by employees. In this way, a strong SCT doesn't weaken PSCs. More precisely, a strong SCT plays an integration role that preserves the differentiation of safety culture produced by the plurality of PSCs.

4 Towards the Construction of "Hybrid" Professionals?

The tension between PSCs and SCT echoes the recent debate about "professionalism" versus "managerialism" in public administrations (Olakivi & Niska, 2017, p. 20):

> Typically, professionals are presumed to resist managerial, economic and governmental requirements as alien intrusions on their professional autonomy (Noordegraaf, 2015). In recent academic debates, however, the image of resistance (e.g. Doolin, 2002) has made room for the notion of "hybridity" (see Noordegraaf, 2015). Instead of resisting managerial intrusions, professionals are seen to balance (Teelken, 2015) or navigate (Croft, Currie, & Lockett, 2015) between managerial and professional imperatives, objectives, interests and requirements (also Reay & Hinings, 2009; Denis Ferlie & van Gestel, 2015)

Olakivi and Niska (2017) suggest that professionalism and managerialism can be interpreted as two "overlapping discourses" in any professional work and organizational action. Such a combination of discourse would produce the emergence of "hybrid professionalism" (Noordegraaf, 2015) (i.e. a form of professionalism that complements organizational and managerial objectives).

Our analysis also suggests a form of hybrid professionalism, but what we see in the domain of safety culture is not just an overlap of two discourses, rather it is a discussion between various professional PSCs and a managerial SCT, which produces local and situated compromises that are beneficial for safety, without "hybridization": the condition for the dialog is to have very clear, legitimate and differentiated PSCs and a strong and well-designed SCT.

5 Conclusion: Three Conditions for the Management of Safety Cultures in a Pluralist Approach

We suggest that "managing" safety cultures is possible and meaningful when it takes a pluralist approach. This may be possible under three conditions. The first is to cease to demand a compliant approach based on the homogeneous alignment of individual and collective behaviors on a single predefined referential. The second is to establish the legitimacy and the value of strong, local practitioners and professional cultures (PCSCs), rooted in day-to-day practices. The third is to implement management tools (SCT) designed to articulate the diversified and differentiated cultures of practitioners and professional groups that work in high-risk technologies.

References

Antonsen, S. (2009). Safety culture and the issue of power. *Safety Science, 47*(2), 183–191.

Daniellou, F., Simard, M., & Boissières, I. (2010). Facteur Humains et Organisationnels de la sécurité industrielle: un état de l'art. Cahiers de la Sécurité Industrielle, Feb 2010.

Detchessahar, M. (2013). Faire face aux risques psycho-sociaux: quelques éléments d'un management par la discussion. *Négociations, 1,* 57–80.

Lawrence, P., & Lorsch, J. (1967). Differentiation and integration in complex organizations. *Administrative Science Quarterly, 12,* 1–30.

Noordegraaf, M. (2015). Hybrid professionalism and beyond: (New) Forms of public in changing organizational and societal contexts. *Journal of Professions and Organization, 2,* 187–206.

Olakivi, A., & Niska, M. (2017). Rethinking managerialism in professional work: From competing logics to overlapping discourses. *Journal of Professions and Organization, 4,* 20–35.

Rocha, R., Mollo, V., & Daniellou, F. (2015). Work debate spaces: A tool for developing a participatory safety management. *Applied Ergonomics, 46,* 107–114.

Star, S. L., & Griesemer, J. R. (1989). Institutional ecology, translations' and boundary objects: Amateurs and professionals in Berkeley's museum of vertebrate zoology, 1907–39. *Social Studies of Science, 19*(3), 387–420.

Tillement, S., Cholez, C., & Reverdy, T. (2009). Assessing organizational resilience: An interactionist approach. *M@n@gement, 12*(4), 230–264.

Weick, K. E. (1987). Organizational culture as a source of high reliability. *California Management Review, 29*(2), 112–127.

Chapter 7
Culture as Choice

David Marx

Abstract Culture is not about outcome, nor about human error. Culture is choice, framed by shared values and beliefs. Creating a strong safety culture means helping employees make good, safe choices. To do that, we must first clearly articulate to our teams both the mission, and the many values we work to protect. For safety, we need to let our employees know where safety fits into the mix, both in theory, and in real world role modeling. Next, we must design our systems and processes to facilitate the choices we want to see. Human choices are somewhat predictable—meaning the system design process can anticipate and resolve impending conflicts before we introduce system or procedural changes. Culture requires work: the everyday task of role modeling, mentoring, and coaching in a manner so that our employees understand how they are to make choices around the value of safety, given a world of conflict between the mission and the many disparate values we hold. And lastly, we need to have the systems in place to monitor our performance. Are we making choices that are supportive of our shared values?

Keywords Culture · Choice · System design · At-risk behavior

1 The Link Between Culture and Harm

Accident. Pilot Error. Medical Error. Mechanical Failure. Employee Mistake.

These are all familiar terms in the safety space. Aircraft accident; pilot error. Patient harmed; medication error. Culture itself is rarely identified in the press as the root cause of public harm. Human error and inadvertency are the hallmarks of our collective 'safety dialogue.'

Yet, once in a while we see an individual who chooses to crash an aircraft, or who chooses to kill a patient. These we quickly distinguish. We call them 'intentional' and claim these acts are outside the purview of 'safety.' These are more a

D. Marx (✉)
Outcome Engenuity, Plano, TX, USA
e-mail: dmarx@outcome-eng.com

C. Gilbert et al. (eds.), *Safety Cultures, Safety Models*,
SpringerBriefs in Safety Management,
https://doi.org/10.1007/978-3-319-95129-4_7

security, or criminal matter; they are not 'safety.' In the United States, once it's determined that an aircraft pilot 'intended' to crash the airplane, the investigation is transferred from the National Transportation Safety Board (the safety investigators) to the Federal Bureau of Investigation (the criminal investigators) (NTSB, n.d.). This dialogue forces us into an uncomfortable and illogical place—that there are only two forms of human behavior: human error, and its evil twin, 'intentionality.' In fact, humans and their behaviors are much more nuanced than these two labels can encompass.

Most corporate adverse events have their origin in two places:

1. the systems we design around the humans, and
2. the choices of humans within those systems.

The resulting harm itself, and the human errors (slips, lapses, and mistakes) that may have caused the harm, are really two forms of outcome—outcomes to be monitored, studied, and perhaps, grieved. Systems and choices are where the action is, with culture referring to the choices made within the system.

The first origin of adverse events is system design. Systems develop over time. From simple surgical instruments to robotic surgery, from messages delivered via horseback to satellite phones, systems keep getting smarter and smarter. Collections of components, physical and human, keep getting more complex and tightly coupled, from getting steam locomotives to run on time, to organizing a mission to Mars. As system designs mature, we try to make the fit right for human beings within those systems. We do our best to design around the inescapable fallibility of human beings—that propensity to do other than what we intended. Better human factors design means less human error.

Choice is the second origin of adverse events. Design safe systems, and help employees make safe choices within those systems. Humans are not computers—we have free will (although this is sometimes challenged amongst experts in the safety field). We make choices that impact the rate of adverse events. That said, understanding human choice is messy business, often set aside in favor of the more simplistic explanation of human error. Even graduate safety courses spend little time on managing choice—it's all about human error. A commercial truck driver who crosses the centerline of a highway may very well be said to have made a human error. Yet, both design of the highway and truck may have contributed to the error, as would natural elements like rain or glare. So too would the choices of the driver contribute to his own error—from the decision to send a text message while driving to the decision to drink and drive.

The measure of culture, whether it's customer or employee safety, privacy (for a hospital), profit, or winning culture (for a sports team), resides within the choices of those within the context or value being discussed. Whether it is the City of New York wrestling with the problem of eight million people deviating from basic traffic laws, or a small manufacturer wrestling with personal protective equipment (PPE) compliance, culture is best thought of as the collective choices of those within the system.

2 Culture: What It's Not

What is culture not? Culture is *not* human error. We are all inescapably fallible human beings. The fact that we make mistakes is just part of the human experience. We can reduce the rate by designing good systems around human beings. But we cannot totally eliminate mistakes, simply because there are so many opportunities to make mistakes. Similarly, culture is not outcome. The fact that one group of physicians might have a higher misdiagnosis rate does not necessarily mean they have a weaker safety culture. There are other factors, from patient acuity to the design of the healthcare systems, that might lead to a higher rate of misdiagnosis.

Sitting in a restaurant, we might hear a tray of glasses break when they hit the floor. For the most part, we'd assume an unfortunate human error led to the undesired outcome of broken glasses. We'd make no inferences about restaurant culture based solely upon a human error and its undesired outcome. Yet, if we walk into a restaurant and see open unclean tables, and a number of employees standing idle, we might wonder why they weren't cleaning the tables when it appears they have the time to do so. This may lead us to think about the culture within that restaurant. We'd wonder about their service standards, or the general cleanliness of the restaurant. We might find ourselves talking about it once seated. It would shape our view of the restaurant as a whole, in a way the inadvertently dropped glasses would not. We might see the failure to rapidly clean the tables as a reflection of their service culture. It does not explain the behavior, but it simply recognizes that the apparent choice of the staff not to clean the tables is somehow reflective of the overall culture of the restaurant.

Culture is, in general, not a reflection of highly culpable or even criminal behavior. Every organization will have outlying behavior on the job, from theft to assault. In the framework of a Just Culture, these choices involve 'knowledge or purpose' toward the harm being caused (Outcome Engenuity, 2016). In the United States in 2016, a total of 5300 Wells Fargo employees were fired for creating unauthorized customer accounts (Egan, 2016). The incentive for the employees involved? Bonuses based upon the number of new accounts created. Given there were 5300 employees involved, could the unauthorized accounts be described as part of the 'culture' at Wells Fargo? The CEO of the institution was pushed out, in large part for his failure to effectively manage his team. Given its widespread occurrence, theft might very well be seen as part of their culture. That said, highly culpable actions tend to be statistical outliers more than any reflection of corporate culture.

There is, in the Just Culture model, a zone of behavior less culpable than knowledge or purpose called 'recklessness' (Outcome Engenuity, 2016). This is not the intention to cause harm, but rather a 'gambling' with unreasonable risk. A driver who texts and drives might be seen as gambling with the lives of others on the road. What makes it 'reckless' is the determination that the driver recognized the risk as both substantial and unjustifiable, but chose to text and drive because it benefitted him in some way. Reckless, by definition, is not a choice to harm. Rather, it is a choice to gamble, to knowingly take a substantial and unjustifiable risk (ibid.). In the manufacturing environment, this might be to climb to dangerous heights without safety gear, simply to save time. If employees know they are taking an unjustifiable risk, that behavior might be deemed reckless.

Recklessness, along with knowledge and purpose to harm, are generally the conduct of outliers within the organization. They are commonly addressed through a formal process of corrective or disciplinary action. Outliers will always exist. They are not, however, the core of culture.

3 Culture as At-Risk Behavior

'At-risk' behavior is the conduct where individuals or groups engage in a risky choice not knowing or incorrectly justifying the behavior as being safe (ibid.). This might be a group of drivers who routinely fail to indicate a lane change, or a group of nurses who routinely fail to wash their hands walking into a patient's room. 'Drift' is a word appropriate to describe at-risk behavior. There are many reasons for behavioral drift. The human does not easily see the hazard to be avoided by adherence to a safety rule. Or, the incentives in the system encourage deviation from a safety rule in order to meet a production objective. It is the presence of 'at-risk' behavior that is the best indicator of what we call 'safety culture.'

As humans, we will exhibit collective choices around particular values. Aviation is known to be a relatively safe endeavor for a passenger, and a 'highly reliable organization' to experts in the safety space (Stralen, n.d.). Yet, it is a dangerous place to work for employees, with a higher lost-workday injury rate than coal miners and commercial fishermen (BLS, 2017). Is it the inherent danger of the work environment that makes the difference? Is it the system design that makes aviation much safer for passengers than employees? Or is it culture, the collective choices of airline employees, that makes the difference?

Culture can be seen as the characterization of a group's collective choices. A safety culture is one where the value of safety is strongly supported. A profit centric culture is one where profit maximization is strongly supported. For a military unit, mission may be the dominant value, even when it means putting a service member in harm's way. If a group's choices are generally aligned with protecting safety, we'd say they have a strong safety culture. If they are not, if there is at-risk behavior throughout the organization, we'd say they have a weak safety culture. This characterization does nothing to solve the problem, but merely suggests that the system is not working as intended. Employees have drifted into risky choices, and it's threatening a value held by the organization, or society as a whole.

If at-risk behavior is the marker of what we call 'culture,' it is independent of whether those behaviors led, on any day, to no harm, minor harm, or a major accident. Those of us who don't walk around the back of our car before getting in will likely never back over an unseen child. These events are rare, making the prevention of harm seemingly tolerant of the at-risk behavior of not walking around the car before backing up. That said, in the U.S. alone, automobile drivers inadvertently back over 2500 kids each year, killing 100 of them (Kids & Cars, n.d.). A workplace (or individual) safety culture may be 'poor' in the sense that the choices of employees are statistically linked to a higher rate of undesired outcomes.

If we are pursuing highly reliable outcomes, choices matter, even when we humans do not necessarily see the hazard attached to non-compliance. Culture can be seen as the degree to which human beings will, through their choices, be protective of a shared value. This often appears as the 'extra effort' it takes to act in protection of a value in the face of a belief that potential harm is uncertain, delayed, or will simply happen to someone else. For example, in the U.S., hospital acquired infections account for 100,000 lost lives a year (CDC, 2016). The number one thing that can be done to prevent these infections is for hospital employees to wash their hands going in and out of a patient's room. Yet, most hospitals have been working for decades to get their compliance rates to even 90% (McGuckin, Waterman, & Govednik, 2009). Hospitals continuously train their employees, redesign soap and alcohol rub dispensers, and make hand hygiene a discussion in daily huddles. All that said, hand hygiene takes extra effort for physicians and nurses, adding roughly 30s to the time in a patient's room, multiplied by thousands of patients over the course of a career. How willing hospital employees are to perform this task is one marker of a hospital's overall safety culture.

4 The Importance of Why

There are views within the academic community that culture is more than choice. In this view, culture is more a description of values and beliefs. There is no reason to challenge this view. The values and beliefs of employees within an organization surely impact their choices; but it is not only values and beliefs that impact choice. We go to a tennis match and we are quiet; we go to a soccer match and we are loud. For most of us, there is no deeply held value or belief that tennis matches should be quiet and soccer matches loud. It is custom, tradition, or culture. We remain silent at the tennis match because others are silent, and because we'll face some admonition from those nearby if we choose to scream. Likewise, if we remain silent during the thrilling parts of the soccer match, a fellow fan might suggest we get on our feet and start to yell like the rest of the crowd. Sometimes, it is simply fear of being different that causes us to behave in a particular way. The choice to remain silent at the tennis match may have nothing to do with our personal values and beliefs. We may actually be wondering why others do not cheer for their favorite player. Yes, values and beliefs are important, but they are not the only factors impacting group choices.

For every risky choice, there is a unique set of factors that come into play. It is oversimplification to suggest that all unsafe choices emanate from some shared set of values and beliefs. Those with the task of creating safe behaviors are well advised to try to understand *why* employees drift into the risky choice. In some cases, the unsafe behavior might occur simply because the employee does not agree that the safety rule is important enough to follow. In other cases, the root of unsafe choice may come from decades of values and beliefs, such as a male pilot who might choose not to communicate critical safety information to a female copilot. It could also be the case that employees choose a behavior simply to avoid sanction. Healthcare privacy laws were enacted with tough sanctions for those healthcare providers who go into a patient's record when there is no clinical reason to be there (HIPAA, 2013). Within hospitals, we saw the policies shift, as well as behaviors. Did the values and beliefs of healthcare providers change overnight? No. It took time; and for a few diehard voyeuristic staff, those values and beliefs never changed. Just as humans gawk (slow down) as they pass by an accident on the road, the desire to see into a movie star's patient record did not likely shift much through the creation of privacy laws. Did the culture change? Yes, if the culture is what we do. No, if the culture is seen as values and beliefs.

5 Improving Culture

As managers and systems designers, we can influence culture. Engage a loud buzzer in a car when a seatbelt is not latched, and drivers will indeed buckle up more frequently. We can shape the choices of human beings, at least at a statistical level. The entire criminal justice system is based upon this premise, as is every human resource policy within an organization (Florida Government). Humans make choices; system designers are out to influence the choices they make, just as marketing companies are out to influence which laundry detergent we buy.

In healthcare, organizations are working hard to create learning cultures where employees can self-report their errors for the purpose of organizational learning. For most hospital staff, this behavior is very much aligned with their individual value of protecting the safety of their patients. Yet many, if not most, employees report only what they cannot hide. The U.S. Agency for Healthcare Research and Quality's Patient Safety Survey routinely finds that fear of punishment is the reason most don't report errors or near misses (AHRQ, n.d.). This is called a 'punitive' culture, not because punishment is among the shared values of the staff, but because employees believe that organizational leaders see punishment as a reasonable tool for controlling staff errors. The failure of employees to report errors and hazards is real. The cause is either apathy that nothing will change, or fear that they will be punished for bringing risks to light.

To be effective managers, we should recognize that human beings are, at our core, hazard and threat avoiders. We speed on the road. We see the speed limit sign, which represents the rule, and we keep going. We see a police car parked up ahead, and we slow down. The police car represents an immediate threat; the speed limit sign does not. Yet, in the organizational space, we write safety rules with the expectation that human beings will somehow blindly follow the rules simply because they are safety rules. When human beings inevitably drift, we claim 'poor safety culture.' A recent U.S. governmental report on an aircraft accident characterized the offending organization as having a 'culture of complacency' (Loreno, 2016). It's easy to attach the label of poor culture; it's a bit harder to understand how mission-oriented employees are reacting to the world around them.

In order to shift culture, to shift choices, it is good to know the reasons behind the behavioral drift. If a task is hard to perform, or gets in the way of the mission, an employee might feel pushed toward non-compliance. If deviation from a safety rule is easy, or if deviation optimizes the mission, an employee might feel the pull of non-compliance. This is particularly true where employees have a hard time connecting the desired safety behavior to the undesired outcome. Many safety behaviors are obvious to the employee involved. Wearing eye protection when using a grinder makes sense because the risk of non-compliance is obvious to employees. Yet, when events become increasingly rare, humans will soon recognize that non-compliance often yields no undesired outcome. We see others deviate from the safety rule, with no bad outcome. Consider our collective inattentiveness to listening to the pre-flight briefing, in large part because we believe that it is unlikely we will ever need to use those safety instructions. We humans shed load we do not see as essential to largely mission-focused work. We ignore the safety briefing on the airplane simply because we want to get on with reading the magazine in our hands.

Top managers have a large influence on culture. By role modeling, mentoring, and coaching their direct reports, they drive the commitment the organization has toward protecting a value like safety. Conversely, top managers can kill a strong safety culture by their actions. Maybe it's the CEO of a railroad who wants to drive the train when he is unqualified, or a director of a manufacturing facility who chooses not to wear a safety helmet and glasses when required. Top managers set the expectation of safety, and through their behaviors, model what a culture of safety looks like.

In order for organizations to improve their safety culture, leaders must be willing to take the lead. They must role model, mentor, and coach their direct reports in a manner that says a little extra effort is worth it. They must be continually cognizant of the role of system design in shaping behavior. They must be cognizant of external cultural norms slipping into the organization, from hierarchical traditions to perceived gender roles.

Line managers must do the same. They must be role models, mentors, and coaches in a manner demonstrating that the extra effort is worth it. The mission never goes away—every employee has production goals. Yet, every organization can and should let its employees know what it means to be protective of a shared value; from putting on protective gear, to taking the time to lock and tag out electrical systems that might endanger an employee.

Culture is not easy because we humans are complex. We are goal oriented. We pursue our missions with zeal, and we find creative ways to do this even when faced with fewer resources, and less time. Cutting corners to get things done is part of the human spirit. Across human endeavors, we shed what we see as the unnecessary rules and guidance mandated by those in control. Even academics have the problem of staying within font size rules when presenting their findings, because they believe the ability to present more is more important than presenting in a legible manner.

It's just who we are. And that's why culture is so hard. In a strong safety culture, the group will hold each other accountable for conforming to the behaviors that support safety. This will hold true even in the face of the generally held belief that potential harm is uncertain, delayed, or will simply happen to someone else. Safety is about preventing harm. Safety culture is about choice.

6 Tangible Steps

Creating a strong safety culture means helping employees make good, safe choices. To do that, we first must clearly articulate to our teams both the mission, and the many values we work to protect. For safety, we need to let our employees know where safety fits into the mix, both in theory, and in real world role modeling. Next, we must design our systems and processes to facilitate the choices we want to see. Human choices are somewhat predictable—meaning the system design process can anticipate and resolve conflicts before we introduce system or procedural changes. After that, we are left with the everyday task of role modeling, mentoring, and coaching so that our employees understand how they are to make choices around the safety value, given a world of conflict between the mission and the many disparate values we hold. And lastly, we need to have systems in place to monitor our performance. Are we making choices that are supportive of our shared values?

References

AHRQ. (n.d.). *Surveys on patient safety culture*™. Retrieved from Agency for Healthcare Research and Quality: www.ahrq.gov/professionals/quality-patient-safety/patientsafetyculture/.

BLS. (2017). *Injuries, illnesses, and fatalities*. Retrieved from Bureau of Labor Statistics: https://www.bls.gov/iif/oshsum.htm.

CDC. (2016). *Healthcare-associated infections*. Retrieved from Center for Disease Control and Prevention: https://www.cdc.gov/hai/surveillance/.

Egan, M. (2016, September 8). 5,300 Wells Fargo employees fired over 2 million phony accounts. *CNN money*. Retrieved from http://money.cnn.com/2016/09/08/investing/wells-fargo-created-phony-accounts-bank-fees/.

Florida Goverment. (n.d.). Chapter 775-012, General purposes. In F. Goverment, *Florida statutes*. Retrieved from Official Internet Site of the Florida Legislature: http://www.leg.state.fl.us/statutes/index.cfm?App_mode=Display_Statute&Search_String=&URL=0700-0799/0775/Sections/0775.012.html.

HIPAA. (2013, October 18). *The reality of HIPAA violations and enforcement*. Retrieved from Health Insurance Portability and Accountability Act: https://www.hipaa.com/the-reality-of-hipaa-violations-and-enforcement/.

Kids & Cars. (n.d.). *Backovers*. Retrieved from Kids and Cars: http://www.kidsandcars.org/how-kids-get-hurt/backovers/.

Loreno, D. (2016, October 18). NTSB announces results of investigation into deadly Akron plane crash. *FOX 8 Cleveland*. Retrieved from http://fox8.com/2016/10/18/ntsb-to-announce-results-of-investigation-into-deadly-akron-plane-crash/.

McGuckin, M., Waterman, R., & Govednik, J. (2009). Hand hygiene compliance rates in the United States–a one-year multicenter collaboration using product/volume usage measurement and feedback. *College of population health faculty papers 48*. Retrieved from http://jdc.jefferson.edu/healthpolicyfaculty/48/.

NTSB. (n.d.). *The investigative process*. Retrieved from National Transportation Safety Board: https://www.ntsb.gov/investigations/process/pages/default.aspx.

Outcome Engenuity. (2016). Just Culture Algorithm™ v3.2.

van Stralen, D. (n.d.). *Models of HRO*. Retrieved from High Reliability Organizing: http://high-reliability.org/High-Reliability-Organizations.

Chapter 8
Safety, Model, Culture

The Visual Side of Safety

Jean-Christophe Le Coze

Abstract In this chapter, I address what I believe to be a complementary discussion for this book on the relationship between safety, models and culture. One interesting angle of analysis is indeed to focus on drawings, graphics or visualisations that have supported powerful heuristics designed to channel ways of thinking the complex topic of safety, analytically and communicatively. In order to build the argument about the importance of how drawings, pictures or visualisation structure the understanding of safety individuals and become a support for action, some illustrations are offered, covering different categories of actors populating high risks systems, from process operators to engineers and managers. From there, a discussion of more research oriented drawings is developed, based on two illustrations: the Heinrich-Bird pyramid and the Swiss Cheese Model. They are considered from several analytical categories including their generic, normative, metaphoric aspects along with their status as inscriptions, boundary and performative objects.

Keywords Safety models · Visualisations · Drawings · Heuristics

1 Safety, Model and Culture

Safety culture, safety model, model of safety culture, maybe even, culture of safety model…how to think about these three words and their relationship? How do safety, culture and model actually relate to each other? The two expressions, safety culture or safety model have been around for a long time, and I believe that the notion of safety model is older than that of safety culture. For instance, Barry Turner, who created the incubation model of disaster, was a sociologist who developed a strong interest for the topic of culture. He published one of the first books in the sociology of organisation on the topic of culture (Turner, 1971).

J.-C. Le Coze (✉)
Ineris, Verneuil-en-Halatte, France
e-mail: Jean-Christophe.LECOZE@ineris.fr

C. Gilbert et al. (eds.), *Safety Cultures, Safety Models*,
SpringerBriefs in Safety Management,
https://doi.org/10.1007/978-3-319-95129-4_8

The idea of safety culture emerged after Chernobyl, in the late 80 s, as described in many writings (e.g., Cox & Flin, 1998). The incubation model of disaster emphasised the need to consider the problem of information handling in organisations, and was very much about the now very popular concept of learning.

James Reason, a psychologist, developed one of the most famous safety models in the field during the 1980s (Reason, 1990a, 1990b) but only later applied the notion of safety culture to the topic (Reason, 1997), with no previous experience, as a psychologist, of using the notion of culture (Reason & Mycielska, 1982). Other prominent authors in the field could be added here to illustrate further the many different uses of these notions, and their relationships. For this reason, I think that the connection between safety, model and culture is a complex one, something that somehow becomes yet more complex when considering the visual aspect of models. The aim of this chapter is therefore to raise awareness among readers about this dimension of models which represents an important aspect of safety practices and research.

2 The Visual Side of Safety

In the past few decades, history, sociology, anthropology and the philosophy of science and technology have engaged in a strong and sustained interest for the world of visualisations (Coopmans, Vertesi, Lynch, & Woolgar, 2014). Latour has been one of the early promoters of this interest for the materialisation of scientific practices (Latour, 1986; Latour & Woolgar, 1979). In "*Thinking with eyes and hands*", Latour convincingly argued for paying greater attention to the diversity of what he described as "inscriptions", these textual, graphic or computerised supports and traces of all sorts that scientists manipulate daily to describe, to approach, to depict, to comprehend, to conceptualise, to explain, to anticipate, to predict phenomena. They provide the concrete visual support through which one can build or construct networks of inscriptions within which an understanding of phenomena is possible, and a world enacted.

Of course, beyond the practice of science, in our current image-saturated culture, including the issue of big data, people rely on a kind of visual literacy which consists of constructing meaning from everything we see. The notion of visual literacy—derived from the notion of literacy initially developed to investigate the relation between thinking and writing (Olson, 1998)—is an ability developed from early age to adulthood in order to evolve and to cope in an environment in which a very good part of our information is received through our eyes.

Translated into the daily practices of the process industry, one can indeed observe that there is a world of images made of texts, signs, diagrams (including PID: process instruments diagrams), alarms, thresholds, schemas, tables, pictograms, posters, procedures, schedules, Gantt charts, indicators, maps, logs, forms, etc., supporting and guiding actions (Le Coze, 2015; Le Coze, forthcoming). These pictures are based on graphical features such as lines, shapes, colours, spaces and textures but also balance, variety, movement, proportion, etc. Here are some examples and selections of pictures, drawings and graphics which structure the environment of safety management in high risk systems. I indicate artefacts which primarily concern the activity of different actors of safety critical systems.

2.1 Control Rooms Interfaces

Interfaces are probably the first visualisations which come to mind because of how much they frame the activities of process operators in control rooms, of pilots in the cockpit, of surgeons in operating theatres, etc. Of course, it has been an important research area in the field of cognitive engineering since the early 1980s (Rasmussen & Lind, 1981), and there are now many established writers and standards publications on the topic (e.g. Bennet & Flach, 2011). The importance of such visualisations is obvious, and an ethnographic study of practices in a control room of a chemical plant shows how much interpretive work by the operators is required to navigate between the (animated and interactive) pictures and the real chemical, physical or electrical processes (Fig. 1).

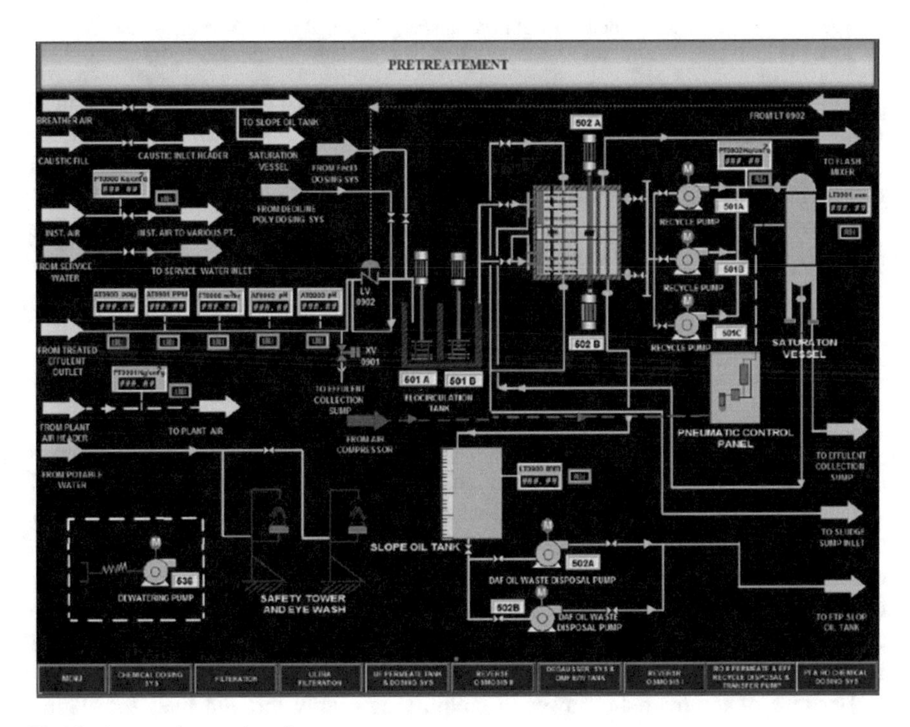

Fig. 1 A control room interface

Examples abound of situations of uncoupling between the information provided by the interfaces, events and people's sense making. Alarms for pilots creating much confusion due to design choices, as in the Rio-Paris case, or sensors in a raffinate tower indicating the wrong level of liquid to operators in the Texas City case, are two illustrations where visual environments shape decision-making processes. It is precisely because of this potential of a discrepancy between what is happening, what is graphically represented and what is constructed in the mind of users of interfaces that great care should be given to visualisations.

2.2 *Risk Assessment Matrices*

Engineers also rely on drawings and visualisation to help assess risks and design safe processes. Analysis by Tufte of the graphics which supported the decision rationale of the Challenger launch in 1986 has become a landmark study of this aspect of engineering decision making (Tufte, 1997). By omitting to exhibit in an appropriate manner data which were available and that they knew to be important to ground their rationale, engineers failed to provide a more complete view of the relationship between temperatures and rings' problems.

> The chart makers had reached the right conclusion. They had the correct theory and they were thinking causally, but they were not displaying causally. (Tufte, 1997, 44)

The pattern of issues with rings as temperatures dropped was not visible for all to see, especially managers who had to be convinced.

Considering that this issue was at the heart of the debate between engineers and managers the night before the launch, the importance of visualising is made convincingly by Tufte, although retrospectively. Viewing data differently could have made a difference in the decision-making process.

2.3 *Safety Trends*

Operators and engineers are, of course, not the only users of graphics; managers also rely on many of them in their activities. The most obvious examples in the field of safety are the trends based on indicators which are built and followed to steer organisations' degree of achievement in preventing health, occupational or process events. The widespread use of ratios in occupational safety calculating the number of days off for injured people per 1000 h worked (at the level of a plant or an entire corporation), but also the number and magnitude of incidents which are considered important to follow, are transformed into graphics.

Because of the heuristic power of simplifying reality through lines translating trends, such visualisations are extremely popular in management circles. They synthetize or aggregate data for quickly grasping trends that need to be supervised and acted upon. As trends rise and fall, managers have to explore causes and to look for explanations in order to maintain or to improve situations that are considered inadequate. Of course, doing so means going beyond what is made available through this reduction of reality through numbers translated in graphics.

2.4 Constructing Safety Through Seeing

Interestingly, our understanding of safety as a construct enacted on a daily basis by a multitude of artefacts, actors and institutions, has never really been seen from the angle of these drawings, pictures or visualisations. Emphasis on cognition, organisation or regulation through established disciplines such as cognitive psychology, sociology of organisation or political sciences has framed our understanding of safety in the past 30 years. Little credence has been given to a transversal appreciation of visual artefacts across descriptions and conceptualisations. However, the sustained attention from science and technology studies (STS) on pictures, diagrams or inscriptions (in a Latourian sense as introduced above) has gradually raised awareness of their importance among safety researchers involved in qualitative empirical case studies.

For instance, the concept of "coordination centres" by Suchman (1997) is more materialistically based than the notions of "heedful interactions" or "collective mindfulness" derived from HRO studies (Weick & Roberts, 1993; Weick, Obstfeld, & Sutcliffe, 1999), indicating this renewed interest by ethnographers in the embedded context of practices, cognition and social networks. So, our understanding of reliability, resilience or safety would gain from a greater attention to how cognition, organisation or regulation are supported by processes of representing graphically, of explaining through drawing, of visualising safety.

2.5 Researching Through Drawing

But, what can be established from ethnographic fieldwork about the practices of personnel in high-risk or safety-critical systems, whether operators, engineers or managers, namely the importance of drawings, pictures or visualisations for interpretation (and action), is in fact no different for the scientists studying these practices. In particular, safety researchers who interact with professionals in organisations also rely greatly on an array of drawings, graphics or visualisations. Conceptual issues such as comparing high-risk systems, framing sociotechnical systems, theorising safety and accidents, representing human error or establishing causality have been supported by drawings, graphics or visualisations (Le Coze, forthcoming). Two examples regarding the problem of theorising safety are now discussed to defend this thesis, the Heinrich-Bird Pyramid (HBP) and the Swiss Cheese Model (SCM).

2.6 *The Heinrich-Bird Pyramid*

The Heinrich or Bird Pyramid (HBP) is a very well-known triangle deconstructed into several layers (Fig. 2). Practitioners have been long keen to use this representation to design prevention strategies. The image is therefore performative, namely it supports action. This kind of pyramid is built on ratio which differs from time to time; according to authors.

Hale studied the rationale behind these representations (Hale, 2001). His conclusion is that the original authors never implied a connection, such as a causal relationship, between minor and major injuries or events.

> What is therefore surprising is how the strong belief came to get established among safety practitioners, and apparently also among researchers that the causes of major and minor injuries are indeed the same. This seems to be an example of an urban myth (…) We are not going to get very far in preventing major chemical industry disasters by encouraging people to hold the handrail when walking down stairs. (Hale, 2001)

Against this prevailing interpretation, Hale (2001) and subsequently Hopkins (2008) offer alternative views to challenge the implied causality between minor and major injuries (Fig. 3).

Safety practitioners are actually often very keen on deconstructing the rationale of Fig. 2 and react very positively to the alternative pyramids. But, as far as I know, the pyramid as presented in Fig. 2 remains very popular and dominant, and these alternatives have yet to be used and disseminated. The pyramid case remains a very simple yet powerful example of the visualisation of safety. On the basis of a visualisation and its interpretation, safety professionals derive preventive actions, revealing the performative character of this visualisation, namely its ability to take part in enacting specific practices.

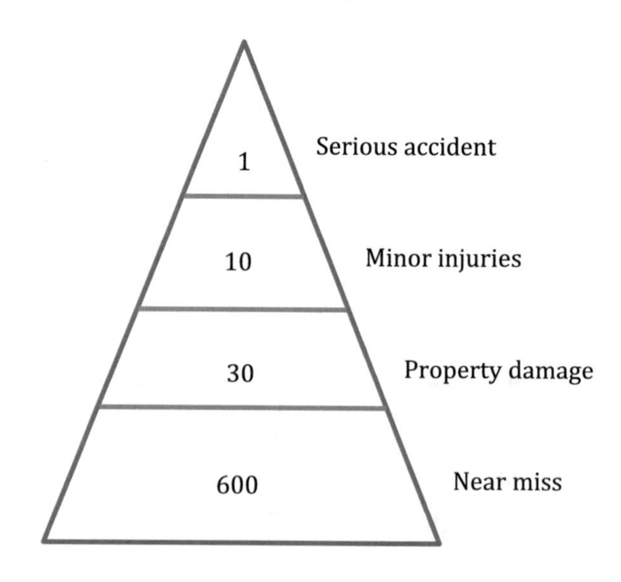

Fig. 2 The Heinrich-Bird pyramid

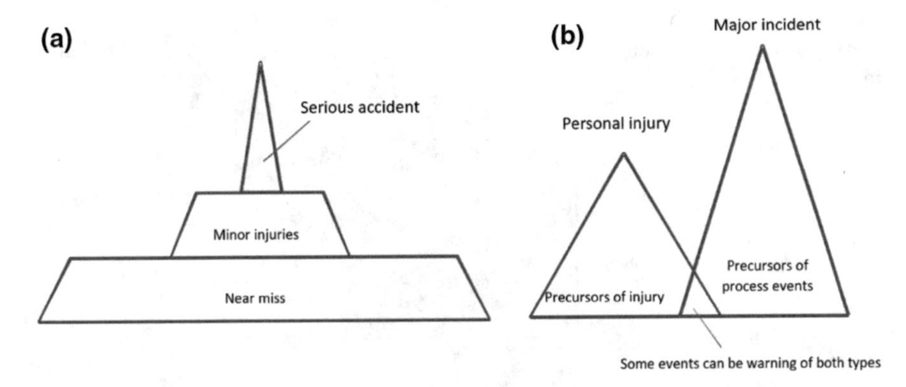

Fig. 3 Hale's (**a**) and Hopkins' (**b**) alternative versions of the Heinrich-Bird pyramid

And part of its success lies indeed in its graphical properties. The fact that Hale and Hopkins created different and alternative options of the pyramid is proof that they, at least implicitly, recognised its visual heuristic strength. One way to "break the spell" is for them to transform the image, to deconstruct its rationale visually transmitted, to show the limitations of the pyramid and what it implicitly conveys. These authors hope to trigger renewed interpretations of the relationship between near misses and major accidents by substituting one representation with another.

The matter is not superficial, accident investigations in the industry have shown that organisations relying on indicators of occupational safety obtaining very good results can suffer major process accidents. This situation therefore directly challenges some of the graphics introduced above and widely used by managers (Fig. 3b). Without careful appreciation of what it is that the lines represent, managers could take for granted, based on the HBP, that occupational safety improvement equals process safety improvement and therefore wrongly interpret the trends.

2.7 A More Sophisticated Example: The Swiss Cheese Model

Some models of safety are much more sophisticated than the HBP (Le Coze, 2013). One example is the popular Swiss Cheese Model (SCM), originally developed in the 1980s (Reason, 1990a) (Fig. 4).

Reason's initial approach (Reason, 1990b) is—to use the words of the author—based on a metaphor and a more workable theory. The starting point of the model is empirical. It consists of the outcomes of accident investigations conducted in the 80s, particularly reports on the King's Cross fire and the Herald of Free Enterprise disaster (Reason, 1990b). These reports enabled him to distinguish between '*active failures*' and '*latent failures*'. For Reason,

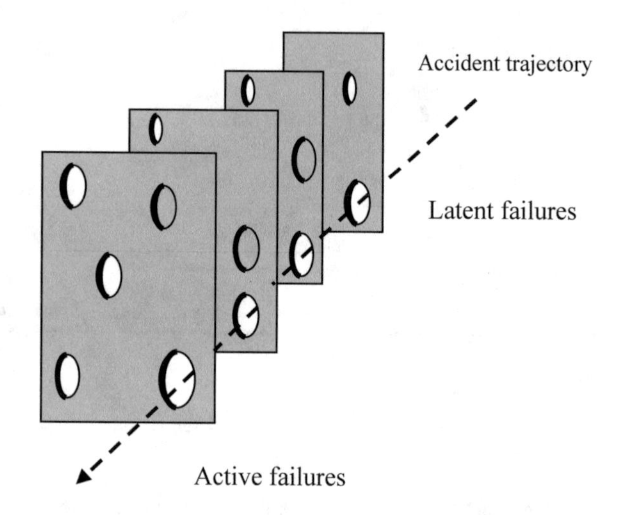

Fig. 4 Defence in depth model (adapted from Reason, 1997)

> Latent failures have their primary systemic origin in the errors of high-level decision makers. (Reason, 1990b, 31)

The metaphor used is a medical one:

> latent failures are analogous to the resident pathogens within the human body, which combine with external factors (stress, toxic agencies, etc.) to bring about disease. (Reason, 1990b, 29)

The 'workable theory,' to go beyond the metaphor, consists of a series of basic elements: decision makers, line management, preconditions, productive activities and finally, defences. They represent a series of 'planes', where

> several factors are required to create a 'trajectory of opportunity' through these multiple defences. (Reason, 1990b, 33)

The success of this model is probably even greater than the HBP.

As commented in Table 1, the success of both models, HBP and SCM, can be explained by several factors (Le Coze, 2016). First is their capacity to be generic, whether in the chemical industry, in aviation or in the railways, both models adapt fairly well (1) and normative because they provide principle for assessing specific situations (2). They also have the ability to mobilise appealing metaphors (3), to be inscriptions (4) and to become boundary objects (5) with a performative dimension (6).

3 Strengths and Weaknesses of Visual Models

Because safety is somehow a product of how these inscriptions participate in structuring and shaping the world of safety practitioners and researchers, these visual models have also triggered a certain number of questions about their

Table 1 H-B Pyramid and SCM properties

Properties	Heinrich-Bird Pyramid (HBP)	Swiss Cheese Model (SCM)
Generic (1)	Can be applied to any systems	
Normative (2)	A certain threshold of near misses indicates a higher likelihood of more serious or major events	Many holes in different slices indicates higher vulnerability
Metaphoric (3)	A pyramid which locates an event to be avoided at the top, and expressed in terms of size of surfaces	A number of slices with a space between them representing defence in depth with holes which can be aligned then traversed by an arrow
Inscriptions (4)	Very much part of the network of actors who interact on a daily basis in the context of safety critical activities	
Boundary objects (5)	Allow these different kind of actors, and sometimes researchers and practitioners, to discuss, to debate then to promote and to design solutions or programs of safety improvement	
Performative (6)	Ability to support action, to transform reality through what they imply in terms of preventive measures to be designed and implemented	

limitations and drawbacks. Just as the HBP was challenged, as shown above (Hale, 2001; Hopkins, 2008), the SCM has been analysed and criticised (Dekker, 2002; Hollnagel, 2004; Reason, Hollnagel, Pariès, 2006; Shorrock, Young, Faulkner, 2004; Turner & Pidgeon, 1997). This certainly demonstrates their popularity and the need to reflect on them.

Now, if considering the HBP from its strengths and weaknesses, one could say that HBP is valuable because it is simple and easily understandable by anyone. It visually expresses a ratio of events (600:30:10:1) coupled with their intensity (from near miss to major) with, at the top of the triangle, a major event to be prevented. The downside to this strength is that it implies a dubious causal relationship and hides some complexities. It conflates all kind of events into broad categories and it supports therefore an intuitive but false belief about how safety is produced. The problem is that it influences preventive strategies on a very large scale.

The case of SCM needs a bit more elaboration because of its higher degree of sophistication. On the positive side, the model provokes an immediate intuitive understanding and provides a very clear similarity with the technical approach of 'defence in depth', which translates very well, metaphorically, from technology to organisation (a). It expresses and reduces the complexity of the problem of accidents by indicating the many potential (but unfortunate, or 'normal') combinations of holes that characterise an accident sequence (b). It allows the user(s) to imagine that there are practical recommendations to be derived from the model, by targeting and improving some selected defences (c). It indicates a distance from the targets (the damages), so that incidents can be expressed by their level of proximity to a catastrophe, and therefore offers, in principle, the possibility of a normative assessment (d). It distinguishes between proximal and remote individuals, who play a different role in the genesis of accidents, and is, in some senses, systemic rather than individualistic in this respect (e).

On the negative side, it does not explain clearly what the holes are in reality—users are left to translate this for themselves—it is only suggestive and not analytical (a′). It does not indicate how holes are likely to align (b′). It relies on an underlying philosophy of failure and errors (whether 'latent' or 'active'), introducing the notion of blame either at the level of proximal or remote actors (c′). It is not explicit or insufficiently specific about the slices or planes (or defences) although they are to be associated with different scientific fields (psychology, management, sociology, etc.) and leaves a lot of room for interpretation about how slices are to be considered (e.g. functions, actors, procedures) and how far the slices should go back in space and time (d′). It offers a linear and sequential view of accident trajectories, as a sequence of events following each other over time, and cannot account for multiple and/or circular causalities with different time spans (e′).

In sum, HBP and SCM are challenged by a number of authors in the field and exhibit indeed a multiplicity of positive and negative aspects. They are two famous examples of safety models for which their visual properties are specifically at the heart of their heuristic value and power to explain, to make sense and to perform (note also that their downside is that they are ways of not seeing, they lock users in certain interpretations). With regards to the intention of this article to justify the relevance of paying attention to the drawings, graphics and visualisations available when one studies safety, it is now, I hope, perfectly clear that they find their place among the more textual models of sociology or the mathematical ones of engineering… .

4 Conclusion

The chapter addresses a complementary topic for this book: the visual side of safety. Safety, model and culture are interwoven notions which are difficult to disentangle. This chapter adds to the complexity of this conundrum with a focus on the drawings, graphics and visualisations which support the daily practices of a multitude of actors in high-risk systems, but which support research and conceptualisation too. With the help of two examples, the Heinrich-Bird pyramid (HBP) and the Swiss Cheese Model (SCM), it is argued that safety practices and research are intrinsically developed on the basis of some form of drawings, graphics and visualisations which play a central role in the performance of safety-critical systems.

References

Bennett, K. B., & Flach, J. M. (2011). *Display and interface design. Subtle science, exact art.* Boca Raton, FL: CRC Press, Taylor & Francis group.

Coopmans, C., Vertesi, J., Lynch, M, E., & Woolgar, S. (2014). *Representation in scientific practice revisited.* Cambridge, MA: MIT Press.

Cox, S., & Flin, R. (1998). Safety culture: Philosopher's stone or man of straw? *Work and Stress, 12,* 189–201.

Dekker, S. W. A. (2002). Reconstructing human contributions to accidents: The new view on error and performance. *Journal of Safety Research, 33,* 371–385.

Hale, A. R. (2001). Conditions of occurrence of major and minor accidents. *Journal of the Institution of Occupational Safety and Health, 5*(1), 7–21.

Hollnagel, E. (2004). *Barriers and prevention.* Aldershot, UK: Ashgate.

Hopkins, A. (2008). *Failure to learn. The BP Texas City refinery disaster.* Sydney: CCH.

Latour, B. (1986). Visualization and cognition: Thinking with eyes and hands. *Knowledge and Society: Studies in the Sociology of Culture Past and Present, 6*(1), 1–40.

Latour, B., & Woolgar, S. (1979). *The social construction of scientific facts. Introduction par Jonas Salk.* London, Beverly Hills: Sage Publications.

Le Coze, J.-C. (2013). New models for new times. An anti dualist move. *Safety Science, 59,* 200–218.

Le Coze, J.-C. (2015). Existe-t-il une pensée graphique en sécurité industrielle? Colloque Réactions à risque. Regards croisés sur la sécurité dans la chimie. Le 9 décembre, Mines ParisTech/INERIS/Centre Max Weber, Paris.

Le Coze, J.-C. (2016). *Trente ans d'accidents. Le nouveau visage des risques sociotechnologiques.* Toulouse: Octarès.

Le Coze, J.-C. (forthcoming). Visualising safety. In J.-C. Le Coze (Ed.), *Safety research: Evolution, challenges and new directions.* London: Routledge.

Olson, R. D. (1998). *L'univers de l'écrit. Comment la culture de l'écrit donne forme à la pensée.* Paris: Retz.

Rasmussen J., & Lind, M. (1981). *Coping with complexity.* Risø Report, Roskilde, Risø National laboratory.

Reason, J. (1990a). *Human error.* Cambridge: Cambridge University Press.

Reason, J. (1990b). The contribution of latent human failures to the breakdown of complex systems. *Philosophical Transactions of the Royal Society of London. Series B, 327,* 475–484.

Reason, J. (1997). *Managing the risk of organizational accidents.* Alderhsot, Hampshire, England: Ashgate.

Reason, J., Hollnagel, R., & Pariès, J. (2006). *Revisiting the 'Swiss Cheese' model of accidents.* EEC Note No. 13/06. Eurocontrol.

Reason, J., & Mycielska, K. (1982). *Absent-minded? The psychology of mental lapses and everyday errors.* Englewood Cliffs, NJ: Prentice Hall.

Shorrock, S., Young, M., & Faulkner, J. (2004). Who moved my (Swiss) cheese? The (R) evolution of human factors in transport safety investigation. In *ISASI 2004 Proceedings.*

Suchman, L. (1997). Center of coordination: A case and some themes. In L. B. Resnick, C. Pontecorvo, & B. Burge (Eds.), *Discourse, tools and reasoning: Essays on situated cognition* (pp. 41–62). Berlin: Springer.

Tufte, E. R. (1997). *Visual explanations. Images and quantities, evidence and narratives.* Connecticut: Graphics Press.

Turner, B. A. (1971). *Exploring the industrial subcultures.* London: Macmillan Press.

Turner, B. A., & Pidgeon, N. (1997). *Man-made disaster. The failure of foresight*. Oxford: Butterworth-Heinmann.

Weick, K., & Roberts, K. (1993). Collective mind in organizations: heedful interrelating on flight decks. *Administrative Science Quaterly, 38,* 357–378.

Weick, K., Sutcliffe, K. M., & Obstfeld, D. (1999). Organising for high reliability: processes of collective mindfullness. *Research in Organisational Behavior, 21,* 81–123.

Chapter 9
On the Importance of Culture for Safety: Bridging Modes of Operation in Adaptive Safety Management

Gudela Grote

Abstract There is no one best way to improving safety performance. Rather, organizations need to have the ability to operate in different organizational modes depending on external and internal conditions. Organizational actors need to recognize and implement switches between modes of operation, e.g. changing from more centralized to more de-centralized work processes and vice versa. It is argued that organizations are confronted with but also actively construct different conditions for safety with respect to the amount of uncertainty they have to manage. Choices about reducing, absorbing, and creating uncertainty along with external demands on the organization require teams to operate in the face of various mixes of stability and flexibility demands. Culture is a strong stabilizing factor, needed particularly when teams have to be very flexible and adaptive. Culture can also help to build the interdisciplinary appreciation required for integrating highly diverse knowledge in search of the most effective solutions to safety problems.

Keywords Uncertainty · Adaptive capacity · Safety culture · Switching operational modes

1 Introduction

A key question in the long-standing debate between proponents of different conceptual approaches to safety is whether there is one best way to achieve it (Grote, 2012). Many well-known safety theories and models would seem to imply that indeed one size fits all, be it the organizational abilities of responding, monitoring, anticipating, and learning in resilience engineering (Hollnagel, Pariès, Woods, & Wreathall, 2011), the reporting, just, flexible, and learning culture advocated by Reason (1997), or the five characteristics of high-reliability organizations (HRO): preoccupation with failure, reluctance to simplify, sensitivity to operations,

---

G. Grote (✉)
ETH Zürich, Zürich, Switzerland
e-mail: ggrote@ethz.ch

C. Gilbert et al. (eds.), *Safety Cultures, Safety Models*,
SpringerBriefs in Safety Management,
https://doi.org/10.1007/978-3-319-95129-4_9

commitment to resilience, and deference to expertise (Weick & Sutcliffe, 2001). However, a closer look at the reasoning for these characteristics reveals that they all build on the fundamental insight that organizations need to be able to switch between different modes of operation in order to respond to changing internal and external demands (LaPorte & Consolini, 1991; Weick & Roberts, 1993).

The requirement for organizations to be adaptive and the specific measures organizations need to take in order to fulfill this requirement are the starting point for the discussion to follow. First, it will be argued that organizations are confronted with but also actively construct different conditions for safety with respect to the amount of uncertainty they have to manage (Grote, 2016). Second, the modes of operation needed to respond to these conditions will be reflected upon with respect to requirements for safety management. Lastly, the role of culture in helping to bridge different modes of operations will be discussed and recommendations for building and maintaining an appropriate culture presented.

2 Approaches to Uncertainty Management

Uncertainty is understood in its most basic form as 'not knowing for sure' due to lack of information and/or ambiguous information (Daft & Lengel, 1984; Galbraith, 1973; ISO 31000, 2009). There is a growing consensus that managing risk and safety not only entails the systematic consideration of quantitative and qualitative uncertainty in risk assessments (e.g., Bjelland & Aven, 2013), but also choices between reducing, absorbing *and* creating uncertainty as part of risk mitigation (Amalberti, 2013; Griffin, Cordery, & Soo, 2016; Grote, 2009, 2015; Pariès, 2016). These choices are influenced by the conditions organizations face with some having to operate in more uncertain environments, e.g. due to strong competition, or having to accomplish tasks which inherently contain more uncertainty, e.g. complex problem solving. Additionally, choices are impacted by requirements for risk control, such as, for instance, those prescribed by regulatory agencies. Moreover, decision-makers in organizations may hold different worldviews regarding adequate risk management, which will also influence preferred approaches to managing uncertainty.

Grote (2015) summarized existing approaches to uncertainty management into three broad categories (see Table 1). *Reducing uncertainty* to a level of acceptable risk, the main thrust in classic risk mitigation, is built on the belief that safety can only be achieved in stable systems with a maximum of central control. This belief favors safety measures such as standardization and automation in order to streamline work processes. *Absorbing uncertainty* comes from acknowledging the limits to reducing uncertainty in complex systems and the corresponding belief that safety stems from a system's resilience, that is its capacity to recover from perturbations. Within this belief system, control is to be decentralized, based for instance on the empowerment of local actors and fast feedback loops. Finally, the importance of *creating uncertainty* is inherent in a worldview that stresses

Table 1 Approaches to uncertainty management

	Reducing uncertainty	Absorbing uncertainty	Creating uncertainty
Objective	Stability	Flexibility	Innovation
Conceptual approach	Classic risk mitigation	Resilience	Self-organization
Control paradigm	Central control	Control by local actors	Shaping contexts for local actors
Examples of safety measures	Standardization; automation	Empowerment; fast feedback loops	Setting constraints for experimentation

Adapted from Grote (2015)

self-organization and innovation as drivers for safety. Local agents are assumed to be controllable only by shaping contexts for their adaptive behavior, for instance through setting incentives and constraints for experimentation.

Carroll (1998) has pointed out that different conceptions of uncertainty management tend to be prevalent in different professional (sub)cultures within organizations. While engineers and executives believe in uncertainty reduction through design and planning, operative personnel are very aware of the need for resilience in the face of only partially-controllable uncertainties. Lastly, social scientists, in their role as consultants or human factor specialists for example, will also argue for openness to learning and innovation, thereby promoting the benefits of uncertainty creation.

3 Different Modes of Operation in Response to Changing Uncertainty Landscapes

Depending on external and internal conditions and the choices made regarding reducing, absorbing or creating uncertainty, organizational actors find themselves confronted with different demands on the stability and flexibility of their behavior (Grote, 2015; Pariès, 2016; Vincent & Amalberti, 2016). The main drivers for seeking stability are demands on predictability, reliability, and efficiency, or more generally on control. These demands are created within organizations, but they may also stem from external sources such as regulatory bodies. Highly dynamic and uncertain environments tend to form the main source of flexibility demands (Thompson, 1967). However, flexibility needs also arise from within the organization due to complex production processes or possibly the opposite—highly routinized work processes, where over-routinization and complacency are to be avoided by introducing variation and change (Gersick & Hackman, 1990).

Table 2 illustrates how organizations in different industry sectors, different functions within organizations and different work tasks may rely on the three options for handling uncertainty. Thus, in organizations which overall are geared

Table 2 Illustration of options for managing uncertainty at different organizational levels

	Reducing uncertainty	Absorbing uncertainty	Creating uncertainty
Industry sector	Nuclear power	Health care	Oil exploration
Organizational function	Production planning	Operations	R&D
Work task	Routine task	Problem-solving	Inventing

Adapted from Grote (2015)

towards reducing uncertainty there will be certain functions and work tasks that require absorbing or creating uncertainty and vice versa. Accordingly, demands on stability and flexibility will vary across different parts of the organization and possibly within single units when work tasks change. Additionally, Vincent and Amalberti (2016) have pointed out that the most effective uncertainty management even for the same work task within the same organizational unit can vary due to changing working conditions, e.g. staff shortage or time of day.

How the actual organizational processes may differ in response to varying demands for stability and flexibility can be illustrated for the coordination within work teams (Grote et al., 2018):

- When both stability and flexibility demands are low, as for instance in team debriefings where the focus is on sharing knowledge and learning outside of acute work pressures, coordination mostly happens among team members without much reliance on formal leadership or organizational rules.
- When stability demands are high and flexibility demands are low, as in many process control tasks, the emphasis is on efficient production, usually enabled by structural coordination mechanisms embedded in technology and standard operating procedures, leaving little need for leadership or mutual adjustment among team members.
- When stability demands are low and flexibility demands high, for instance in teams that have to innovate at all cost, coordination happens by mutual adjustment and shared leadership to bring all team members' competences and resources to bear on idea generation and implementation.
- When stability and flexibility demands are high because both highly reliable performance of complex tasks and fast reactions to unpredictable change are required, a broad range of coordination mechanisms has to be employed in parallel, helping teams to maintain control, e.g. through directive leadership and/ or strong shared norms, and be adaptive, e.g. through sharing leadership tasks.

Teams may have to move quickly between the four conditions and switch their mode of operation accordingly. A surgical team may perform a routine operation (high stability, low flexibility) followed by a complex emergency operation (high stability, high flexibility). It will also undertake team debriefings (low stability, low flexibility) and may engage in experimenting with a new operating technology (low stability, high flexibility). As a consequence, continuous monitoring of stability and flexibility requirements and of necessary adaptations following decisions on reducing, absorbing or creating uncertainties is crucial for the comprehensive management of risk and safety.

An additional distinction to be made in order to define the best possible modes of operation for any given situation is that between personal or occupational safety and process safety. Personal safety is related to hazards that can directly damage the worker's health and well-being, such as exposure to toxic substances or mechanical forces. Workers need to protect themselves against these hazards which often creates tasks outside of their primary work task, for instance by having to wear personal protective equipment when repairing high voltage power lines. The second kind of safety is process safety. Here the work process contains risks for others beyond the workers themselves, such as passengers on a train or aircraft, patients being operated on, or people living next to a power plant. Safety requirements are inherent to the performance of these work processes and do not create extra tasks for the workers involved.

Personal and process safety may be related to different parts of work processes and may or may not coincide for workers and other affected individuals. During an operation, a surgeon handles process risks for his or her patient which do not contain personal safety issues for him or herself. However, the risk of infection exists for both the patient and the surgeon. An interesting example to illustrate this distinction is hand hygiene. Health care personnel wear gloves to protect both the patient and themselves. However, depending on which function is salient for them, they will be more or less careful about touching non-sterile objects with their covered hands (Jang et al., 2010).

Demands on personal safety tend to be predictable and the required behavior is prescribed in safety rules and monitored by the team itself and by supervisors and auditors. Accordingly, personal safety can be said to increase demands on the stability of team behavior. Demands on process safety also increase stability demands, especially when the level of risk embedded in the work process is very high. However, inasmuch as work processes are complex and only partially predictable due to high levels of external or internal uncertainties, process safety needs to be ensured by concurrently responding to high flexibility and high stability demands. As discussed earlier, flexibility demands may not only be imposed on the team, but also wilfully created in search of innovative solutions to problems and opportunities for learning. These are the situations that according to Perrow's (1984) seminal analysis are unmanageable because organizations are ill-equipped to handle concurrent centralization and decentralization demands stemming from tightly coupled and highly complex processes. Seeking ways to manage these situations has motivated much of the research on fostering resilience and adaptive capabilities in teams and organizations.

Griffin et al. (2016) in their summary of research on organizational adaptive capabilities required for adequate safety management have gone a significant step further still. They argue that there is not only a need to be adaptive in day-to-day operations, but also in response to demands for major organizational change. This "dynamic safety capability" includes three components (Griffin et al., 2016, p. 254):

- sensing, which refers to the ability to scan and interpret the external environment for opportunities and threats to safety;
- seizing, which refers to the ability to integrate complexity by managing contradictions and competing goals related to safety;
- transforming, which refers to second-order change aimed at modifying core safety capabilities and transforming processes and procedures.

4 The Role of Culture for Adaptive Safety Management

The requirement to manage multiple organizational forms and to help organizational actors switch between them in response to changing external and internal demands is broadly discussed in the management and organizational literatures, especially under the headings of managing paradox (Smith & Lewis, 2011) and organizational ambidexterity (O'Reilly & Tushman, 2013). For the most part, the role of culture is touched upon in very generic terms in this research, pointing to the necessity to build common norms and values that help bridge apparent contradictions such as discipline and stretch, control and flexibility, or diversity and shared vision (Gibson & Birkinshaw, 2004; Lewis & Smith, 2014; Wang & Rafiq, 2014). Some authors argue that ambidexterity—the ability to concurrently exploit existing knowledge and to explore new ideas—can only be achieved by having dedicated organizational units operating in flexible versus stable modes supported by the respective cultural mindsets. They stress the crucial role of senior management who have to create an overarching vision, while also communicating the need for resolving the inevitable trade-offs and conflicts inherent in organizational ambidexterity (O'Reilly & Tushman, 2008).

With respect to safety management, the role of culture as a source of adaptive change has not yet received systematic attention, as stated by Griffin et al. (2016). One attempt to describe linkages between culture and adaptive safety management has been made by Reiman, Rollenhagen, Pietikäinen, and Heikkilä (2015). Based on the literature on complex adaptive systems, they outline a number of tensions similar to those already mentioned (e.g., trade-offs between repeatability and flexibility or between global and local goals) and argue that a more mature safety culture will develop when these tensions are explicitly addressed. Another approach to capturing the contribution of culture to organizational adaptiveness has been to define certain core values which should be shared, foremost mindfulness (Weick & Sutcliffe, 2001), that is the readiness to continuously scrutinize existing and emerging expectations within a larger context. A mindful culture, or to use Reason's (1997) terms, an informed culture, contains four components: reporting culture, just culture, flexible culture, and learning culture. The latter two components in particular refer to an organization's adaptive capabilities, supporting, for instance, sensing mechanisms in teams that allow them to recognize changing environmental demands and switch modes of operation accordingly.

Where does all of this leave a dedicated senior manager keen to develop adaptive capacity and the cultural basis to support that adaptive capacity in his or her organization? Three general recommendations can be derived from existing research, which will be described below.

4.1 Recommendation 1: Understand the Limits to Managing Culture

As has been stated most prominently by Schein (1992), organizational culture comprises patterns of shared basic assumptions that groups develop as they learn to cope with internal and external challenges in their organization and that are taught to newcomers in the organization as the correct way to see the world. From this definition follows that cultural change is usually slow and not fully predictable. Culture is affected by safety management measures as through all other activities in the organization, but this process cannot be centrally managed nor prescribed. Culture generally shows itself most clearly during organizational change when basic assumptions are challenged. Therefore, instead of prescribing a certain kind of culture, senior managers should be alert to any indication of resistance to change towards more safety, aim to identify specific cultural norms and assumptions that may be the source of this resistance, and work towards changing those norms and assumptions.

Attempts to assess culture are generally only meaningful if they can serve as leading indicators of safety, that is if they help to identify norms and assumptions that potentially hurt safety performance. When used as part of post hoc explanations for accidents and incidents, culture tends to obscure the picture because, by focusing attention on very broad assessments of norms and values, it distracts from manifest organizational and management problems. An example is the expert report on the BP Texas City accident (Baker, 2007), where many problems in the work organization were mentioned, but not analyzed in much detail, only to conclude that inadequate safety culture was a major cause of the accident. Due to the inherent difficulties in observing culture and in evaluating what a "good" culture is, organizations are best advised to assess safety management rather than culture. Shared perceptions of safety management, which are captured by safety climate questionnaires (Flin, Mearns, O'Connor, & Bryden, 2000) have been shown to be a valid leading indicator for safety performance (Christian, Bradley, Wallace, & Burke, 2009).

4.2 Recommendation 2: Foster Culture as a Stabilizing Force in Adaptive Organizations

Culture itself is a coordination mechanism, which helps to integrate work processes and build a shared understanding of work goals and means to achieve them. Thereby culture serves as a 'soft' centralization mechanism for decentralized operations in organizations. As Weick (1987, p. 124) has described it:

> (Culture) creates a homogeneous set of assumptions and decision premises which, when they are invoked on a local and decentralized basis, preserve coordination and centralization. Most important, when centralization occurs via decision premises and assumptions, compliance occurs without surveillance.

Shared basic assumptions encapsulated in organizational and team culture are a crucial stabilizing mechanism for otherwise highly adaptive behavior including switches between different modes of operation (Grote, 2007). For instance, a shared norm of always speaking up with concerns and ideas will better help to master unexpected challenges than any attempt to cover all the possible turns situations can take by means of standard operation procedures (Grote, 2015). Another example is psychological safety, which refers to the shared belief that it is safe to take interpersonal risks in a team (Edmondson, 1999). Psychological safety acts as a stabilizing factor in teams, freeing resources for handling the substantial cognitive demands arising from highly-uncertain situations.

4.3 Recommendation 3: Build Culture by Bridging Worldviews and Accepting Ambiguity

Building an overarching culture of interdisciplinary appreciation (Grote, in press) is crucial for bridging the worldviews embedded in the different approaches to uncertainty. Adaptive safety management depends on a shared understanding across professional boundaries of the legitimacy of reducing, absorbing and creating uncertainty in response to complex and dynamic situations. This can be achieved by promoting perspective taking and cross-learning among the different professions involved in safety. The diverse belief systems have to be reflected on and sufficiently reconciled to create shared views on problems and on ways to solve them.

Acknowledging different perspectives on problems and possible solutions also results in a high tolerance for ambiguity. Rather than declaring one perspective as being correct, decision-makers have to balance different perspectives and make difficult trade-offs. This also holds for leaders more generally who cannot follow one best way of leading, but have to have a broad portfolio of behaviors at hand to answer to changing stability and flexibility demands. Formal leaders may have to step back to let team members do the leading at one moment and may have to resume control in a directive fashion shortly after if conditions change fast (Klein,

Ziegert, Knight, & Xiao, 2006). The importance of this dynamic capability has long been recognized in the management literature (e.g., Denison, Hooijberg, & Quinn, 1995), but acquiring it in practice remains a challenge.

5 Final Remarks

The main argument in this chapter has been that organizations need adaptive safety management in order to make adequate choices between reducing, absorbing, and creating uncertainty and to support teams in changing their modes of operation in response to those choices as well as external conditions. Beyond building the mindful or informed culture that is generally considered a solid foundation for adaptive safety management, the fundamental role of culture as a powerful stabilizing force that helps to coordinate action and integrate work processes in decentralized and flexible modes of operations should be taken into account and employed wisely. Regarding the particular nature of cultures that are beneficial for adaptive safety management, one crucial aspect is respect for the viability of different perspectives on problems and their solutions. Such a culture of interdisciplinary appreciation is at the heart of bringing all knowledge in organizations to bear on finding the most effective ways to promote safety.

References

Amalberti, R. (2013). *Navigating safety*. Dordrecht: Springer.

Baker, J., (2007). *The report of the BP US refineries independent safety review panel.*

Bjelland, H., & Aven, T. (2013). Treatment of uncertainty in risk assessments in the Rogfast road tunnel project. *Safety Science, 55,* 34–44.

Carroll, J. S. (1998). Organizational learning activities in high-hazard industries: The logics underlying self-analysis. *Journal of Management Studies, 35*(6), 699–717.

Christian, M. S., Bradley, J. C., Wallace, J. C., & Burke, M. J. (2009). Workplace safety: A meta-analysis of the roles of person and situation factors. *Journal of Applied Psychology, 94* (5), 1103–1127.

Daft, R. L., & Lengel, R. H. (1984). Information richness: A new approach to managerial behavior and organizational design. In L. L. Cummings & B. M. Staw (Eds.), *Research in organizational behavior* (Vol. 6, pp. 191–233). Homewood, IL: JAI Press.

Denison, D. R., Hooijberg, R., & Quinn, R. E. (1995). Paradox and performance: Toward a theory of behavioral complexity in managerial leadership. *Organization Science, 6*(5), 524–540.

Edmondson, A. (1999). Psychological safety and learning behavior in work teams. *Administrative Science Quarterly, 44*(2), 350–383.

Flin, R., Mearns, K., O'Connor, P., & Bryden, R. (2000). Measuring safety climate: Identifying the common features. *Safety Science, 34*(1), 177–192.

Galbraith, J. (1973). *Designing complex organizations*. Reading, MA: Addison-Wesley.

Gersick, C., & Hackman, J. R. (1990). Habitual routines in task-performing groups. *Organizational Behavior and Human Decision Processes, 47*(1), 65–97.

Gibson, C. B., & Birkinshaw, J. (2004). The antecedents, consequences, and mediating role of organizational ambidexterity. *Academy of Management Journal, 47*(2), 209–226.

Griffin, M. A., Cordery, J., & Soo, C. (2016). Dynamic safety capability: How organizations proactively change core safety systems. *Organizational Psychology Review, 6*(3), 248–272.

Grote, G. (2007). Understanding and assessing safety culture through the lens of organizational management of uncertainty. *Safety Science, 45*(6), 637–652.

Grote, G. (2009). *Management of uncertainty—Theory and application in the design of systems and organizations*. London: Springer.

Grote, G. (2012). Safety management in different high-risk domains—All the same? *Safety Science, 50*(10), 1983–1992.

Grote, G. (2015). Promoting safety by increasing uncertainty—Implications for risk management. *Safety Science, 71,* 71–79.

Grote, G. (2016). Managing uncertainty in high risk environments. In S. Clarke, T. Probst, F. Guldenmund, & J. Passmore (Eds.), *The Wiley-Blackwell handbook of the psychology of occupational safety and workplace health* (pp. 485–505). Chichester, UK: Wiley.

Grote, G. (In press). Social science for safety: Steps towards establishing a culture of interdisciplinary appreciation. In *Human and Organizational Aspects of Assuring Nuclear Safety—Exploring 30 Years of Safety Culture*, Proceedings of an International Conference organized by IAEA, Vienna, February 2016.

Grote, G., Kolbe, M., & Waller, M. J. (2018). The dual nature of adaptive coordination in teams: Balancing demands for flexibility and stability. Accepted for publication in Organizational Psychology Review.

Hollnagel, E., Pariès, J., Woods, D. D., & Wreathall, J. (2011). *Resilience engineering in practice: A guidebook.* Burlington, VT: Ashgate.

ISO 31000. (2009). *Risk management—Principles and guidelines*. Geneva: ISO.

Jang, J.-H., Wu, S., Kirzner, D., et al. (2010). Focus group study of hand hygiene practice among healthcare workers in a teaching hospital in Toronto, Canada. *Infection Control and Hospital Epidemiology, 31*(02), 144–150.

Klein, K. J., Ziegert, J. C., Knight, A. P., & Xiao, Y. (2006). Dynamic delegation: Shared, hierarchical, and deindividualized leadership in extreme action teams. *Administrative Science Quarterly, 51*(4), 590–621.

LaPorte, T., & Consolini, P. M. (1991). Working in practice but not in theory: Theoretical challenge of "High Reliability-Organizations". *Journal of Public Administration Research and Theory, 1*(1), 19–47.

Lewis, M. W., & Smith, W. K. (2014). Paradox as a metatheoretical perspective: Sharpening the focus and widening the scope. *Journal of Applied Behavioral Science, 50*(2), 127–149.

O'Reilly, C. A., & Tushman, M. L. (2008). Ambidexterity as a dynamic capability: Resolving the innovator's dilemma. *Research in Organizational Behavior, 28,* 185–206.

O'Reilly, C. A., & Tushman, M. L. (2013). Organizational ambidexterity: Past, present, and future. *Academy of Management Perspectives, 27*(4), 324–338.

Pariès, J. (2016). *Comparing HROs and RE in the light of safety management systems.* Unpublished manuscript.

Perrow, C. (1984). *Normal accidents: Living with high risk systems*. New York, NY: Basic Books.

Reason, J. T. (1997). *Managing the risks of organizational accidents*. Aldershot, UK: Ashgate.

Reiman, T., Rollenhagen, C., Pietikäinen, E., & Heikkilä, J. (2015). Principles of adaptive management in complex safety–critical organizations. *Safety Science, 71,* 80–92.

Schein, E. H. (1992). *Organizational culture and leadership*. San Francisco: Jossey-Bass.

Smith, W. K., & Lewis, M. W. (2011). Toward a theory of paradox: A dynamic equilibrium model of organizing. *Academy of Management Review, 36*(2), 381–403.

Thompson, J. D. (1967). *Organizations in action*. New York: McGraw-Hill.

Vincent, C., & Amalberti, R. (2016). *Safer healthcare: Strategies for the real world.* Cham, Switzerland: Springer.

Wang, C. L., & Rafiq, M. (2014). Ambidextrous organizational culture, contextual ambidexterity and new product innovation: A comparative study of UK and Chinese high-tech firms. *British Journal of Management, 25*(1), 58–76.
Weick, K. E. (1987). Organizational culture as a source of high reliability. *California Management Review, 29*(2), 112–127.
Weick, K. E., & Roberts, K. H. (1993). Collective mind in organizations: Heedful interrelating on flight decks. *Administrative Science Quarterly, 38,* 357–381.
Weick, K. E., & Sutcliffe, K. (2001). *Managing the unexpected.* San Francisco: Jossey-Bass.

Chapter 10
Safety Culture and Models: "Regime Change"

Mathilde Bourrier

Abstract The goal of this chapter is to explore the generic organizational challenges faced by any high-risk organization and how they shape the social production of safety. Confronted with six generic categories of challenging dilemmas, high-risk organizations differ in their organizational responses, and in the mitigation strategies they put in place. However, this diversity does not mean that there is an infinite number of options. In the chapter, we introduce the concept of "safety regimes", as a way to tackle the diverse ways in which companies operate, hence leaving aside the somewhat overused "safety culture" concept. The notion of "regime", understood as a stable enough organizational equilibrium, offers an alternative way of documenting the organizational responses that high-risk organizations choose to develop and their direct or indirect consequences for the production of safety. The conditions for devoting such attention to the quality of organizing cannot be prescribed and decided upon once and for all. Rather than proposing top-down safety culture programs, and trying to make them fit into an ever-diverse and surprising reality on the ground, this chapter looks at another analytical option: clarifying the key dimensions that are fundamental to the establishment and comparison of safety regimes.

Keywords High-risk organizations · Safety culture · Culture for safety
Safety models · Safety regime

1 Introduction

As a socio-anthropologist of organizations I have conducted ethnographic studies by comparing similar organizations (nuclear power plants) in a same sector (civil nuclear industry), confronted with similar problems in different countries, regions

M. Bourrier (✉)
Department of Sociology and Institute of Sociological Research,
University of Geneva, Geneva, Switzerland
e-mail: mathilde.bourrier@unige.ch

C. Gilbert et al. (eds.), *Safety Cultures, Safety Models*,
SpringerBriefs in Safety Management,
https://doi.org/10.1007/978-3-319-95129-4_10

and contexts (France; the U.S). I have also compared organizations in different sectors (nuclear industry, global health, health providing institutions, and more recently a police department), dealing with sensitive activities, in different countries (France, Switzerland, Japan, the U.S). This cross-national and cross-sectoral perspective using ethnographic studies led me to focus on generic organizational dynamics, which concern a wide spectrum of high-risk organizations, where safety is paramount.

Rather than proposing a cultural model of high-risk organizations, highly dependent on cultural influence, difficult to measure and empirically unfounded (Bourrier, 2005), I choose to consider the challenges that these peculiar organizations have to face daily and compare their organizational responses and mitigation strategies. As Rochlin observed:

> The challenge is to gain a better understanding of the interactive dynamics of action and agency in these and similar organizations and the means by which they are created and maintained. These arise as much from interpersonal and intergroup interaction as from more commonly studied interactions with external designers and regulators. The interaction is a social construction anchored in cultural dynamics, therefore there are wide variances in its manifestation even for similar plants in roughly similar settings. (Rochlin, 1999: 1558)

The same approach applies to safety models, often presented as gold standards. Current theories in use, from "HRO", "Resilience Engineering" to "Sensemaking" and "Ultra Safe Systems", have been understood too often as prescriptive labels. Deviations from their theoretical principles would signal difficulties in achieving consistent delivery of safe performance. However, most of these theories are descriptive by nature. Even though they are different in their propositions and should not be considered as simple substitutes, they still belong to a certain category of theoretical attempts, post-TMI, aiming at making sense of "surprises in the field" (i.e. the complex and largely unpredictable interactions between technology and humans in very demanding systems). Despite their differences and debates, these theories all seek to move away from a stereotyped view of daily operations and, more essentially, to produce a much more complex and rich view of the social production of safety (for a nuanced presentation of these theories and their respective traditions, see Le Coze, 2016).

These schools of thought propose a useful categorization that combines a number of organizational features and promise to improve the management of high-risk organizations and allow continuous improvement (Bourrier, 2011; Le Coze, 2016). However, their respective list of properties[1] has never been envisioned by their creators as a definitive list. Understood as a process, safety is essentially a never-ending organizational learning process, a "dynamic non-event" in Weick's

---

[1]For HRO theorists, these properties include "self-adaptive features of networks", "having the bubble", "heedful interactions", "attention to failures and learning", "socializing processes emphasizing safety", and "Migration Decision Making". For the resilience school, other angles are of interest: "reliability of cognition and resilience"; "Situation awareness and expertise" (Naturalistic Decision Making); "System safety and accident models adaptation", "Self-organization and complexity" as well as the 4 cornerstones: Anticipating, Monitoring, Responding and Learning.

own words. Paradoxically, and for reasons that remain to be uncovered, these theories, however well-known, have not easily travelled to the shop floor level. Concrete examples of operationalization at the plant level are still lacking (Le Coze, 2016).

To add some complexity to this already dense picture, it is also my observation that most of the time, each high-risk organization offers a unique response to monitoring their intrinsic challenges and displays only parts of the so-called "safety models". Hybridization, borrowings, innovative mitigation strategies, and local adaptations are also central to the social production of safety. This is where the notion of "safety regime" offers an alternative for documenting organizational trade-offs that high-risk organizations develop. This includes the examination of their consequences (direct or indirect) for working teams and for the social production of safety.

2 "Safety Culture", "Safety Cultures", "Cultures for Safety"

"Safety culture" has been and still is a powerful catchphrase to introduce a wide variety of issues that do not fit easily into many models of risk reduction, risk management or risk governance, either those in place or those currently being developed. "Safety culture" gave a voice to the issue of variance, to the "that's the way we do business here" philosophy and to the irreducible levels of variation, when confronted with the large spectrum of organizations in society today (Perrow, 1991; Scott & Davis, 2015). High-risk organizations constitute one subset, and share characteristics with large technical bureaucracies, or large socio-technical systems.

However, safety culture is largely an "after-the-fact" concept. When reading investigation reports after an accident, one often finds paragraphs stressing the "lack of safety culture", commenting on a "broken safety culture" (CAIB, 2003), or a "silent safety culture" (Rogers, 1986). Nevertheless, the concept has some merit (Guldenmund, 2000), even if it was unrealistic to expect such a wide scope of services from one single concept. One way to overcome the limitations often expressed about the concept is to move from "safety culture", as a set of norms and programs to "cultures for safety", as a set of practices. More than a cosmetic vocabulary twist, this expression might offer space to embrace a diversity of models, responses and options, viable on the ground and debatable in practice and in theory.

Behaviors, actions, practices, beliefs, decisions, opinions, innovations, designs, hypothesis and perceptions are influenced by a variety of factors. They all constitute the fabric of a corporate culture, which simultaneously is influenced by all of the above. Corporate cultures, especially in industries where risk and uncertainty are paramount, include a safety culture component, and produce cultures for safety.

Within high-risk organizations, it is highly probable that the "culture for safety" component is a key component that impacts the rest of the corporate culture. Simultaneously, these companies and their plants are also in business to deliver services and products to clients. Safety cannot absorb and monopolize their entire business. Safety is a pre-condition, not a goal in itself. As scholars pointed out long ago: in companies without safety, there is no socially acceptable business. The "Social License" that they are granted explains why they tend to be heavily regulated and sometimes even over-compliant (Gunningham, Kagan, & Thornton, 2004; La Porte & Thomas, 1995).

The safety culture concept masks the fact that professions, trades and units inside a high-risk organization develop different sets of safety knowledge according to their specific needs and positions within the organization. These sets are a mix of rules (prescribed, rehearsed, shared), norms and gold standards (of the trade), customs ("the way we do things around here"), innovations and brilliant improvisations, when the situation demands it. Some rules are taught during the apprenticeship, some are discovered in situ, on the job, and stabilized collectively, and others rely on personal and intimate know-how, which an expert masters over time.

Hence, different safety cultures/cultures for safety coexist in an organization. Safety culture/culture for safety is therefore a synthesis of generic convictions and the management's visions, intertwined with professional norms and each professional's experience. It presents itself more as a dialectic than as a set of principles to be followed from A to Z. The key message here is that in various parts of the organization, safety is produced, thought through, and sometimes bargained about. Safety is the object of intense trade-offs between concurrent objectives: regulatory demands, budget constraints, production pressures, planning constraints, technical innovations, workforce fatigue, to name but a few.

Rochlin (1993) and Moricot (2001) once argued that "friction" between trades, professions and units had something to offer for the sake of safety. The conditions under which such a sometimes controversial debate can be supported cannot be prescribed and decided upon once and for all. Diversity of safety cultures is often seen as an obstacle, whereas it could well be interpreted as a sign of a mature and dynamic organization.

3 On the Limited Usage of "Safety Models" at the Shop Floor Level

A lot of insight and inspiration has been provided by theories originating in the eighties and nineties, such as "HRO", "Resilience", "Sensemaking", and "Ultra-Safe Systems". Their creators (La Porte, Roberts, Rochlin, Schulman, Hollnagel, Levenson, Weick, Sutcliffe, Amalberti…) sought to explain how and why some organizations did better, were safer, learned better, than others. How to explain the variability? What could account for the differences? What was transferable from one industry to another?

Their detailed descriptions, empirical fieldwork, and conceptual propositions nourished and still nourish scholars across disciplines and industries (Grabowski & Roberts, 2016; Haavik, Antonsen, Rosness, & Hale, 2016; Le Coze, 2016). However, one is compelled to observe that little has filtered back to the industry level. A lot of what has been discovered and put on the table has mainly served as labelling opportunities for some companies: "We—at such and such company—are an HRO"). There has been a lack of traction at the workplace level to implement some of the insights present in the pioneering works.

We offer one possible (partial) explanation for this delay in the practical application of concepts from these safety models (Bourrier, 2017): using them as gold standards failed to address the complex situations encountered on the ground. Their powerful inspirational capacity has been misused, by freezing the options and reducing the complexity, rather than revealing and opening up the subtle trade-offs that may deserve reinforcement or constructive criticism. Paradoxically, though their initial intention was to make sense of a much more diverse picture than what "traditional" safety science studies were offering (for example, a much more complex understanding of human error and decision-making or cognition), their contemporary usage has led to a reduced set of properties. Their level of abstraction does not always help to understand the pressing dilemmas that workers and teams face daily, nor help interpret their corresponding local adaptations.

In the next section, we examine how we came to think of "safety regimes" as a way to escape from the current stalemate.

4 Introducing "Safety Regimes"

The social production of safety is the result of complex transactions and is influenced by several factors, ranging from demographics of the workforce, type of labor relations, economic and financial resources at hand, regulatory frameworks, technical options and innovation. In this approach, daily work activities for example are central to the production of safety, but they evolve and develop in a precise institutional context, in a specific hierarchical structure, and they espouse an official division of tasks and responsibilities, which in turn creates incentives as well as deterrence.

Evidently, organization theory has long established (Scott & Davis, 2015) that this formal set of structural constraints, no matter how carefully designed, does not tell the whole story of organizational life: rules and formal structure are by nature incomplete and subject to local adaptations willingly worked out by concerned actors. Hence, safety is the product of diverse forces converging sometimes in stable equilibria and sometimes in more vulnerable ones.

Safety can also be approached through the lenses of the "regime" concept, which we have freely borrowed from other fields. In International Relations, *regimes* are commonly defined as

> sets of implicit or explicit principles, norms, rules, and decision-making procedures around which actors' expectations converge in a given area of international relations (Krasner, 1985, p. 2).

This concept taken from political science is fruitful for our discussion. It should not be confused with a regime in the political sense like a *fascist regime* or a *police regime*. Our intent is to capture some features of safety management that depend on the alignment and the hybridization of key organizational characteristics.

The regime concept is used by political economists when they characterize modern Capitalism and its models (Amable, 2003), and by public health experts to characterize the driving forces in the field of Global Health. For example, Lakoff (2010) distinguishes between a "global health security" regime and a "humanitarian biomedicine" regime. They represent two distinct faces, philosophies and rationales for action, in the field of global health, in opposition and sometimes in coordination. He argues that these two regimes give different responses to a set of essential and common issues for global health actors, pertaining to:

1. type of threat;
2. source of pathogenicity;
3. type of organizations and actors;
4. type of techno-political interventions;
5. target of intervention;
6. ethical stance.

Closer to our subject, scholars in Law and Safety science like Baram, Lindøe, and Braut (2013) use the term "Risk Regulatory Regime" to distinguish Norway and the US with respect to their way of regulating off-shore oil and gas companies. They have systematically compared the Norwegian Continental Shelf and US Outer Continental Shelf through five features:

1. legal framework;
2. cost-benefit analysis methodologies;
3. legal standards;
4. inspections and sanctions, and
5. involvement of the workforce.

Inspired by these attempts, our intention is first to identify the key dimensions—that we call also "dilemmas"—lying at the core of how a high-risk organization works and how the organizational responses given to these dilemmas largely shape the "safety regimes". A given "safety regime" produces a certain organizational culture, which in our view encompasses a safety culture. These "safety regimes" have a certain quality of consistency and stability, but are not perfect constructions designed to stay immutable. They are dynamic and their evolution remains partly uncertain and undetermined.

Evolutions affecting these regimes sometimes go unnoticed. They may even lead to misalignment, which might pave the way for "latent human failure conditions" (Reason, 1990), "normalization of deviance" (Vaughan, 1997), or "drift into failure" (Dekker, 2011). In this view, safety culture is a sort of limited proxy which helps understand the type of safety regime one is dealing with. However, studying safety culture by itself and for itself is unlikely to be fruitful. We argue that it needs to be complemented by a thorough investigation of the safety regime in place. This cannot be successfully done without an understanding of the organizational and inter-organizational dynamics at play.

The next section focuses on presenting six categories of problems that structure the type of safety regime which will eventually emerge. Our intention in the rest of this chapter is to clarify the key dimensions that are fundamental to safety regimes.

5 Six Crucial Dimensions

As far as high-risk organizations are concerned, the following issues have to be managed and constantly re-assessed (La Porte, 1996). They constitute complex organizational challenges in all sectors, including the nuclear industry, healthcare, public health interventions and policing. Contingent organizational responses to these dilemmas structure the nature of safety regimes.

First issue: Working with rules and procedures and expanding bodies of regulations is a given in most contemporary organizations (Graeber, 2015), and even more so within high-risk organizations. Yet what are the implications for daily operations of having to document each and every stage of their process?

At a certain level, one could consider that one of the key (cultural) features of these organizations where safety is paramount, is strong and ever-increasing reliance on procedures. The ever-growing scope of proceduralization seems endless and unavoidable (Bieder & Bourrier, 2013). As a consequence, the following crucial questions are constantly on the table: Which groups are in charge of creating, updating rules and procedures, and how is it done? How is the "classic" tension between "prescription" and "autonomy" organized? Ergonomists, human factors specialists along with work sociologists have long demonstrated that procedures are important, but they are imperfect and sometimes counterproductive, when the situation calls for the creation of ad hoc solutions. Also of utmost relevance is how best to maintain a questioning attitude towards rules and procedures when compliant behaviors are most of the time rewarded and sought after. Furthermore, how is the coherence of diverse sets of rules both maintained and challenged?

To these crucial questions, there is no single answer (Bourrier, 1999a, 1999b). For example, for HRO theorists along with Sutcliffe and Weick's observations (2007), one way of mitigating these challenges is to defer to experience. This is called "migrating decision-making", which means that the expert on the ground, closest to the problem, decides on the proper and immediate course of action. At the same time, leaders remain responsible for these decisions, even if in retrospect they are found to be less than adequate. Along with a preoccupation with failure and a sensitivity to operations, deference to expertise ensures that the people who are directly concerned with a problem end up being in a position where they offer their views, their opinions and more importantly their own solutions. But as we know, many organizations favor a different perspective and prefer strict compliance and hierarchical decision-making.

Second issue: How to best plan, schedule, and anticipate activities and at the same time stay alert in order to avoid the complacency that such a planning culture inevitably produces?

This second feature leads us to the core of the culture of these industries: a culture of preparedness, through anticipation, planning and scheduling. Sennett (1998) once brilliantly explained why routine was important in the workplace. Preparation is key, especially in hostile work environments, where risks are present and should be reduced to the minimum possible. Yet this very culture of preparedness, also present in other professional sectors such as public health (Nelson, Lurie, & Wasserman, 2007) is also an identified obstacle to learning how to face the unexpected. To tackle this limitation, HRO theorists and their followers suggest resisting the tendency to simplify.

Often, safety culture programs tend to oversimplify the complex interactions that workers face in their daily jobs, leaving them dubious about the real benefit of such high-level programs which do not capture the rich details of their daily efforts. The false promise of scenario planning has already been documented by Clarke (1999) and is currently constantly re-assessed in the light of special crises and catastrophes, ranging from pandemic influenza A (H1N1) to Fukushima. It should be recalled here, that preparing, scheduling and planning is an important component of any organization dealing with complex operational conditions. However, these tasks should not be understood as distinct phases, but rather as integral to the job while constantly taking into account their limitations.

Weick argued in a seminal article (1987) that storytelling inside high-risk industries was crucial to allow for the circulation and sharing not only of problematic events and how they unfolded, but also of inventive and resourceful options to solve tricky problems. This commitment to resilience is often quoted as an HRO principle. Hollnagel and his colleagues (Hollnagel, Woods, & Levenson, 2006) have the same position when they argue that not enough is shared about resilient strategies. Sharing not only what went wrong, but also what was correctly mitigated is of crucial importance to restoring stability. The tendency of these types of organizations to focus on problems masks the fact that they also have important and unknown resources to leverage some tensions. The situation encountered by Fukushima-Daïchi operators and their management should not be forgotten: sometimes, nothing holds, and capabilities to improvise have to be mobilized by terrified actors left in the dark, resorting to their own meager devices (Guarnieri, Travadel, Martin, Portelli, & Afrouss, 2015; Kadota, 2014).

Third issue: How to best cope with the uncertainties and risks inherent to their process and their institutional environment and, at the same time, maintain products and services at a socially and economically reasonable cost?

Because risks and uncertainties are not known once and for all (they can pile up: e.g. Fukushima), cultivating vigilance for unforeseen combinations of events is important. This can only be done through the constant challenge of rules and procedures in order to interrogate their intrinsic limitations. HRO theorists noticed that a balance has to be established between relying too much on very obedient people, with no inclination to challenge their working environment, and encouraging hotheads, who are always ready to break rules and procedures to set their own norms of performance: HROs do not look for heroes, but for questioning minds.

Despite decades of emphasizing the importance of a "questioning attitude" (INSAG 4) to any safety culture program, striking a balance between compliance and improvisation is probably one of the most difficult issues in people management today.

Fourth issue: How to best enable cooperation among various units, crafts, trades and the intervention of many contractors in order to avoid a silo culture and the formation of clans? How best to depend on highly skilled employees, where almost no substitution is possible among themselves (hence no rotating option), while not getting trapped in their worldviews, entrenched wars and corporatist interests?

As Roberts and Rousseau (1989: 132) explain, with the example of aircraft-carriers in mind, these work environments display a

> hyper-complexity (due to) an extreme variety of components, systems, and levels and tight coupling (due to) reciprocal interdependence across many units and levels.

Anyone interested in organizational design is challenged by this characteristic: the degree of specialization and expertise among the different units, departments, crafts and contractors inevitably provokes a difficulty in communicating each other's concerns. Work is done in silos (Perin, 2005), "structural secrecy" prevails and knowledge does not travel easily throughout the hierarchical structure (Vaughan, 1997). Pockets of crucial knowledge (tacit, formal or informal) can stay hidden for a long time and it does not percolate to the relevant teams or people easily. Vaughan has eloquently demonstrated how the culture at NASA, which relies so much on hard data supported by mathematical modelling, had not allowed other types of evidence to be shared and worked on. As a result, pending issues, deemed illegitimate and hence unresolved, were allowed to strongly contribute to the accidents of *Challenger* and *Columbia*. Attention to the circulation of events, stories and narratives is of crucial importance to counter the traps of "structural secrecy".

Therefore, constantly battling against clans, entrenchment and conventional wisdom is vital for maintaining and improving safety performance. A constant effort has to be made to navigate between trades and reconcile their worldviews to hold on to the big picture. This is typical of the role of management in these types of organizations. The same is true when dealing with the widespread level of subcontracting practices in these industries. Subcontracting requires a lot of reorganization within the client company. Organizing a degree of leadership and "follow-ship" (i.e. capacity to accept to be led by others when required[2]) is a task in itself. Finally, maintaining conditions that allow distributed cognition is a constant challenge that needs to be monitored.

Roberts and Rousseau (1989) coined the expression "having the bubble" to best describe this extraordinary capability to be continually aware of events occurring at different levels. Weick calls it sensemaking activities. In the middle of the 1990s he argued that organizational failure and catastrophic events are best understood as the collapse of collective sensemaking (Weick, 1993). "Mindfulness" and "heedful interrelations" are key properties to maintain collective dedication to safe performance (Weick, Sutcliffe, & Obstfeld, 2008). The quality of the organizational attention on organizing processes is central to their discussion.

Fifth issue: How to best organize the control and supervision of each and every work activity and still count on workers' willingness to adapt the rules when they are not applicable? At least three levels of control can be identified in these organizations:

1. workers themselves must perform their first level of control;
2. first line controllers check that work activities present the correct specifications;
3. a third line of control based on documents, ensures that both workers and first line controllers correctly documented the control they performed on the job.

To ensure these various duties, variations exist, depending on a country's legislation, but also on concrete organizational options. Some companies hire contractors to perform first line controls, whereas others entrust teams of in-house dedicated rotating workers to perform such controls on their own colleagues. This heavy presence of controls, no matter which organizational configuration is in place, generates risks of their own that deserve more study: too much supervision and control potentially creates lack of autonomy, complacency and resentment, lack of confidence, lack of ownership, and dilution of responsibilities. Obviously, too little, or inadequate control also has its drawbacks. The question remains: how are these risks and the associated equilibria carefully studied?

Sixth issue: How to best deal with the strict scrutiny of authorities and regulators?

After Chernobyl, it appeared clearly that the impact of regulators, through their demands and recommendations (or absence thereof), is of crucial importance in the life of heavily regulated socio-technical complex systems. When reading *ex post* reports, it is now common to examine the role of regulators. The nature and the quality of the relationships between regulatees and regulators is often identified as a contributing factor leading to the accident. How does the "culture of oversight"

[2]This odd expression has been used by David Nabarro, who served in 2015 as the Secretary-General's Special Envoy on Ebola. He was referring to the necessity that all actors, including NGOs accept to be coordinated on the ground to enhance international response.

affect safety in the end (Wilpert, 2008)? Here again, the heavy hand that some regulators might impose could be detrimental to safety by reducing room to manoeuver. However, their absence from the scene and their remote access to plants could produce a frustrating and unsatisfactory "paper safety".

6 Conclusion: Regime Change

These questions cannot be solved definitively. They require constant reflection and need to be regularly re-assessed. I once advocated for the creation of organizational observatories to be able to follow the concrete trade-offs that are being made during daily operations (Bourrier, 2002). The concrete and contingent organizational answers given to these problems significantly affect the social construction of safety.

There is no uniform response, yet most high-risk organizations face the same challenges (but do not have the same resources to deal with them). What are the design options that these organizations and networks of organizations, can take, have taken, and are projecting to take in the future? The production of safety is a direct by-product of the organizational regime in place, meant to evolve, and being currently more and more distributed and fragmented among networks of organizations and partners. It colors the kind of safety that is contingently produced in the end.

This is why instead of looking for generic "safety culture" programs and models, it is probably time to accept and welcome the fact that there are different "safety regimes" to ultimately produce safety, none of which are perfect. Some regimes are costly, not only financially but in human terms; others favor strict compliance at the expense of ownership; others promote ownership at the expense of problem sharing and disclosure; some regimes promote compliance, and ultimately create apathy among the teams; some regimes promote autonomy and lack transparency, favoring deep pockets of informal undocumented knowledge; by-the-book compliance can lead to a lack of innovation and can lead to complacency; favoring ad hoc solutions and bricolages, however innovative, can also favor cutting-corners strategies, etc. Depending on the strengths and weaknesses of each trade-off, their final combination will, in the end, impact upon the production of safety and shape the intrinsic qualities of the safety regime in place (cf. Fig. 1).

Finally, each regime carries its own possibilities and limitations. Mitigating risks does not mean suppressing them, because this is not possible. However, regularly assessing the concrete responses to the six dilemmas detailed above offers a space to accommodate and welcome the variance one is witnessing throughout industries, companies and plants.

The *Harvard Business Review* in its April 2016 edition titled: "*You can't fix culture, just focus on your business and the rest will follow*". This blunt statement after decades of articles published on the importance of building strong corporate

Issue 1 - Rules

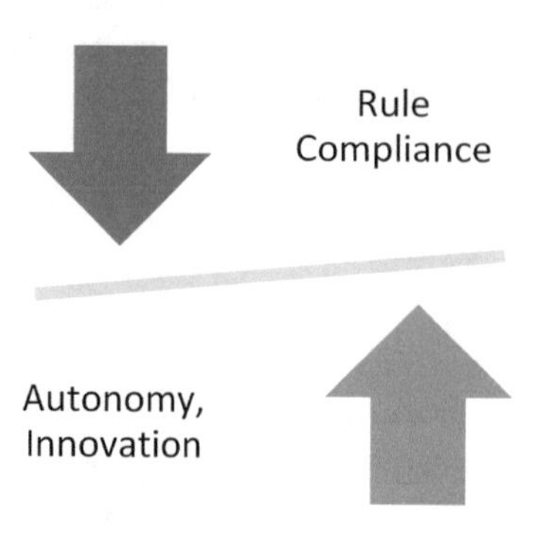

Issue 2 - Preparedness

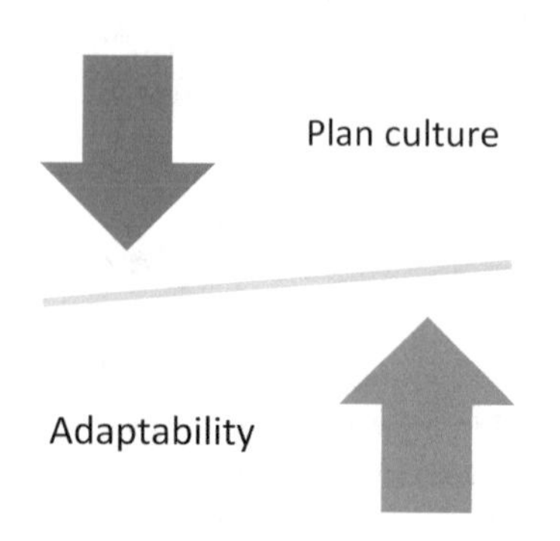

Issue 3 - Uncertainty

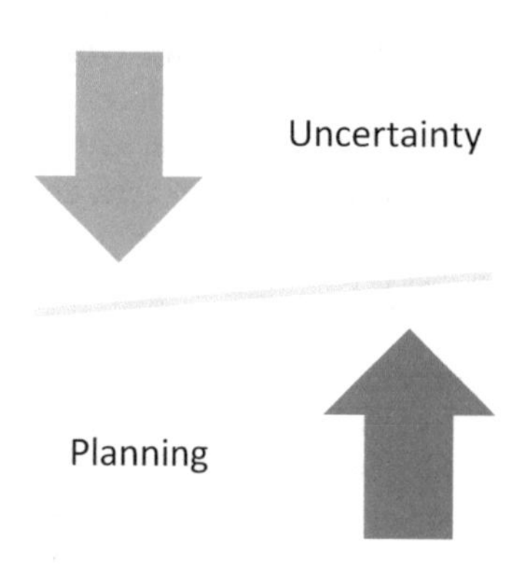

Issue 4 - Specialization

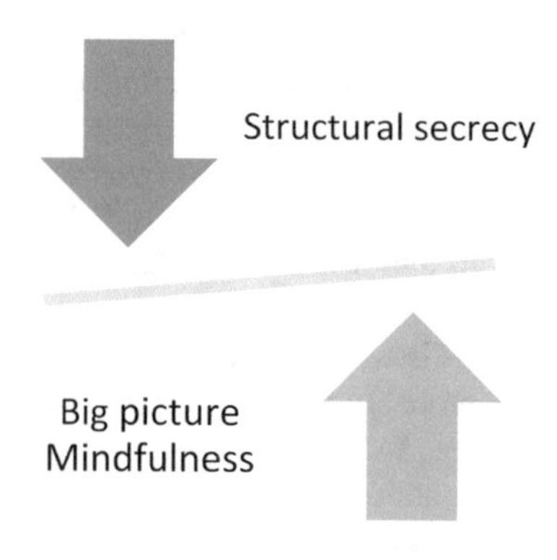

Issue 5 - Control

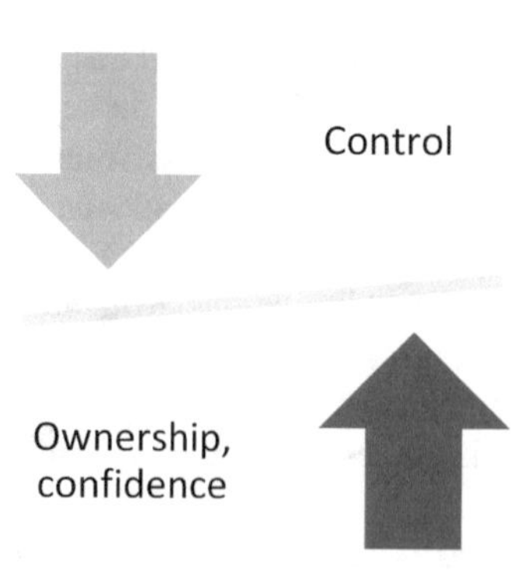

Issue 6 - Regulators' relationships

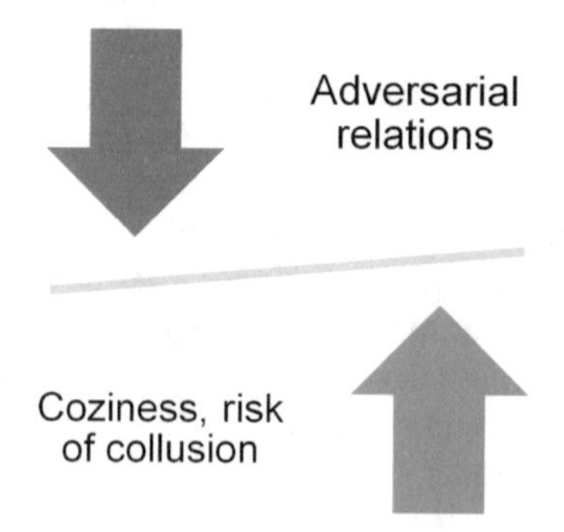

Fig. 1 Trade-offs within safety regimes

culture, promising a thriving business, is somewhat refreshing. The same is probably true with "safety culture": You can't fix safety culture, just focus on your organization and risks and the rest will follow.

References

Amable, B. (2003). *The diversity of modern capitalism*. Oxford: Oxford University Press.

Baram, M., Lindøe, P. H., & Braut, G. S. (2013). Risk regulation and proceduralization: An assessment of Norwegian and US risk regulation in offshore oil and gas industry. In C. Bieder & M. Bourrier (Eds.), *Trapping safety into rules: How desirable or avoidable is proceduralization?* (pp. 69–86). Ashgate-CRC Press.

Bieder, C., & Bourrier, M. (2013). *Trapping safety into rules: How desirable or avoidable is proceduralization?* Ashgate-CRC Press.

Bourrier, M. (1999a). *Le nucléaire à l'épreuve de l'organisation*. Coll. Le Travail Humain. Paris: Presses Universitaires de France.

Bourrier, M. (1999b). Constructing organizational reliability: The problem of embeddedness and duality. In J. Misumi, B. Wilpert, & R. Miller (Eds.), *Nuclear safety: A human factors perspective* (pp. 25–48). London: Taylor & Francis.

Bourrier, M. (2002). Bridging research and practice: The challenge of normal operations studies. *Journal of Contingencies and Crisis Management, 10*(4), 173–180.

Bourrier, M. (2005). L'analyse culturelle: un horizon, pas un point de départ, en réponse à Philippe d'Iribarne. *Revue Française de Sociologie, 46*(1), 171–176.

Bourrier, M. (2011). The legacy of the high reliability organizations project. *Journal of Contingencies and Crisis Management, 19*(1), 9–13.

Bourrier, M. (2017). Organisations et activités à risques: le grand découplage. In J.-M. Barbier & M. Durand (Eds.), *Analyse des activités humaines: Perspective encyclopédique* (pp. 743–774). Paris: Presses Universitaires de France.

CAIB. (2003). *Report of Columbia accident investigation board*. Washington, DC.

Clarke, L. (1999). *Mission improbable: Using fantasy documents to tame disaster*. Chicago: University of Chicago Press.

Dekker, S. (2011). *Drift into failure: From hunting broken components to understanding complex systems*. UK: Ashgate.

Grabowski, M., & Roberts, K. H. (2016). Reliability seeking virtual organizations: Challenges for high reliability organizations and resilience engineering. *Safety Science* (in print).

Graeber, D. (2015). *The utopia of rules: On technology, stupidity, and the secret joys of bureaucracy*. New York: Melville House.

Guarnieri, F., Travadel, S., Martin, C., Portelli, A., & Afrouss, A. (2015). *L'accident de Fukushima Dai-Ichi - Le récit du directeur de la centrale. Volume 1: L'anéantissement*. Paris: Presses des Mines.

Guldenmund, F. W. (2000). The nature of safety culture: A review of theory and research. *Safety Science, 34*(1), 215–257.

Gunningham, N., Kagan, R. A., & Thornton, D. (2004). Social license and environmental protection: Why businesses go beyond compliance. *Law & Social Inquiry, 29*(2), 307–341.

Haavik, T. K., Antonsen, S., Rosness, R., & Hale, A. (2016). HRO and RE: A pragmatic perspective. *Safety Science*, available on line, August 23, 2016.

Hollnagel, E., Woods, D., & Levenson, N. (2006). *Resilience engineering, concepts and precepts*. Farnham: Ashgate.

Kadota, R. (2014). *On the brink: The inside story of Fukushima Daiichi*. Japan: Kurodahan Press.

Krasner, S. (1985). Structural causes and regime consequences: Regimes as intervening variables. In S. Krasner (Ed.), *International regimes*. Ithaca: Cornell University Press.

Lakoff, A. (2010). Two regimes of global health. *Humanity: An International Journal of Human Rights Humanitarianism and Development, 1*(1), 59–79.

Le Coze, J. C. (2016). Vive la diversité! High reliability organisation (HRO) and resilience engineering (RE). *Safety Science*, available on line April 26, 2016.

La Porte, T. R., & Thomas, C. W. (1995). Regulatory compliance and the ethos of quality enhancement: Surprises in nuclear power plant operations. *Journal of Public Administration Research and Theory, 5*(1), 109–137.

La Porte, T. R. (1996). High reliability organizations: Unlikely, demanding and at risk. *Journal of Contingencies and Crisis Management, 4*(2), 60–71.

Moricot, C. (2001). La maintenance des avions: une face cachée du macro-système aéronautique. In M. Bourrier (Ed.), *Organiser la fiabilité* (pp. 161–200). Paris: L'Harmattan.

Nelson, C., Lurie, N., & Wasserman, J. (2007). Assessing public health emergency preparedness: Concepts, tools, and challenges. *Annual Review of Public Health, 28*, 1–18.

Perin, C. (2005). *Shouldering risks: The culture of control in the nuclear power industry* (p. 226). Princeton, NJ: Princeton University Press.

Perrow, C. (1991). A society of organizations. *Theory and Society, 20*(6), 725–762.

Reason, J. (1990). The contribution of latent human failures to the breakdown of complex systems. *Philosophical Transactions of the Royal Society of London Biological Sciences, 327*(1241), 475–484.

Roberts, K. H., & Rousseau, D. M. (1989). Research in nearly failure-free, high-reliability organizations: Having the bubble. *IEEE Transactions on Engineering Management, 36*(2), 132–139.

Rochlin, G. I. (1993). Essential friction: Error control in organizational behavior. In N. Ackerman (Ed.), *The necessity of friction* (pp. 196–234). Springer/Physica-Verlag: Heidelberg.

Rochlin, G. I. (1999). Safe operation as a social construct. *Ergonomics, 42*(11), 1549–1560.

Rogers, W. P. (1986). *Report of the presidential commission on the space shuttle challenger accident.*

Scott, W. R., & Davis, G. F. (2015). *Organizations and organizing: Rational, natural and open systems perspectives*. New York: Routledge.

Sennett, R. (1998). *The corrosion of character: The personal consequences of work in the new capitalism*. New York: W. W. Norton & Company.

Vaughan, D. (1997). *The Challenger launch decision: Risky technology, culture, and deviance at NASA*. Chicago: University of Chicago Press.

Weick, K. E. (1987). Organizational culture as a source of high reliability. *California Management Review, 29*(2), 112–127.

Weick, K. E. (1993). The collapse of sensemaking in organizations: The Mann Gulch disaster. *Administrative Science Quarterly, 38*, 628–652.

Weick, K. E., Sutcliffe, K. M., & Obstfeld, D. (2008). Organizing for high reliability: Processes of collective mindfulness. *Crisis Management, 3*(1), 81–123.

Wilpert, B. (2008). Regulatory styles and their consequences for safety. *Safety Science, 46*(3), 371–375.

Chapter 11
Safety Culture in a Complex Mix of Safety Models: Are We Missing the Point?

Corinne Bieder

Abstract Safety culture is often considered as being the role given to safety in the trade-offs made within an organization. But what is the scope of these trade-offs? If operational activities at the sharp end are naturally included in the safety culture perimeter, other trade-offs are made that structure operational activities, especially through the development of processes, procedures, organizational structure and policies but also through technological choices. These trade-offs are made within the environment of the organization, and that inevitably induces constraints on the role given to safety, as there are already trade-offs inherited from this environment. Likewise, a variety of safety models exist in this environment, in the sense of assumptions or beliefs as to how safety is ensured or more often is to be ensured. Eventually, each organization combines a mix of safety models, some partly conflicting with others. To what extent is an organization aware of the complexity of operations and of what it takes to operate safely? Is this also part of its safety culture? To what extent and how can this complexity be addressed? These are some of the questions addressed in the paper.

Keywords Safety culture · Safety model · Complexity · Trade-offs

1 Introduction

Safety culture as introduced in the early 90s was focused on the importance of the role given to safety compared to other stakes within organizations. Since then, it has been considered a major factor in a safe performance and still receives significant attention in both proactive and reactive safety management. However, it seems that the concept is more commonly used for activities having an obvious and direct impact on operational activities whereas more structuring support functions fall outside of the natural scope of safety culture. Their influence on operational

C. Bieder (✉)
Ecole Nationale de l'Aviation Civile, Toulouse, France
e-mail: corinne.bieder@enac.fr

C. Gilbert et al. (eds.), *Safety Cultures, Safety Models*,
SpringerBriefs in Safety Management,
https://doi.org/10.1007/978-3-319-95129-4_11

activities may be less direct, but is nevertheless significant. Safety culture with its current scope may then become a proxy and prevent a deeper analysis of what actually contributes to safety, especially the high-level decisions sometimes considered rather simplistically as being business decisions only.

Nevertheless, the importance of the role given to safety in trade-offs, whatever the scope of trade-offs considered, may not be the only missing ingredient. Indeed, organizations operate in a context that partly constrains their choices, including the structuring ones, and not only in the role given to safety but also in the way safe performance can be achieved. Eventually, each organization combines a mix of "inherited safety models" throughout its environment, from the authority, its competitors, suppliers… Operating safely or safely enough with this mix of safety models, which are not necessarily consistent with one another, involves not only to give safety an adequate priority but also to navigate the associated complexity.

2 Safety Culture as an Essential Ingredient: The Final Touch or Incorporated All Along?

Safety culture as introduced and defined by the International Atomic Energy Agency (IAEA, 1991),

> "that assembly of characteristics and attitudes in organizations and individuals which establishes that, as an overriding priority, [nuclear power] safety issues receive attention warranted by their significance",

was considered at that time to be the missing ingredient to ensure safety. While other complementary ingredients have been identified since then, following the occurrence of accidents, safety culture often remains an area for improvement on the agenda of many industries. Many definitions have been proposed to try to address the issue, some rather prescriptive, others more descriptive as explained in Chap. 12. If we adopt a descriptive perspective, safety culture can be considered the role or weight given to safety in the trade-offs made within an organization. Yet the scope of the decisions and trade-offs considered is not so obvious. Operational activities at the sharp end are naturally considered within the safety culture perimeter, but what about other activities not directly related to real-time operations such as support functions or high-level managerial decisions? Is safety given sufficient attention in Human Resources policies and decisions or in Procurement policies and decisions?

If we refer to the practical definition of culture proposed by Bower (1966), "the way we do things around here", who does "we" refer to, and what "things" are considered? If normative views of safety culture tend to encompass all the employees of an organization as in safety culture surveys such as that of Eurocontrol (Mearns et al., 2009), descriptive views seem to focus on operators and first-line management, leaving out a whole range of employees and trade-offs not directly involved in real-time operations. Yet, some of these trade-offs and resulting decisions are structuring for the organization and its performance, including the safety performance (Amalberti, 2015; Schein, 2010). A merger or acquisition decision or a transformation of the industrial set-up through increased subcontracting, for example, or a decision to buy new technological devices or new information and management tools are decisions that may significantly impact safety. These decisions, often considered business decisions (as if safety had nothing to do with them, and thus with business), seem to fall outside of the "safety culture" scope despite their role on safety.

Likewise, rules and processes already embed some trade-offs between safety and other business objectives. Whether safety is given sufficient attention in these trade-offs that structure the activities of the organization is not naturally or systematically examined. The intuitive scope of safety culture does not include structuring activities not directly related to real-time operations despite their strong remote influence on them, and thus on safety. For example, a significant number of airlines prescribe in their operations manual to put the autopilot on above 500 or 1000 ft and disconnect it when reaching below this threshold again. Although this rule has an obvious efficiency objective—minimizing the fuel burn or optimizing the use of systems, the impact on safety can be debatable, especially if we consider the long-term impacts on pilots' manual flying skills.

Similarly, the safety impact of new types of pilots' contracts (e.g. pay-to-fly) may not have been analyzed as extensively as their economic impact, without prejudging the conclusions.

In many hazardous activities, some of the structuring choices made at some point in time are later analyzed in terms of their consequences on safety. It is the case of the increasing reliance on subcontractors or of mergers, for example. In the nuclear industry, a specific expert investigation was launched by the French Nuclear Safety Authority (ASN) in 2013 to "examine the efficiency of the measures taken by EDF (Electricité de France) to facilitate the priority given to nuclear safety in the interactions between EDF and the sub-contractor" (ASN, 2013) or of some mergers. In the airline industry, the recent merger between two airlines was mentioned in several accident investigation reports as a factor that may have played a role (Commission of Inquiry into the Air Ontario Crash at Dryden, Ontario (Canada), 1992). Interestingly, while these aspects are analyzed after an accident, thus identified a posteriori as having possibly played a role in the accident, they are still considered a priori "business decisions" and made with limited safety considerations, if any. Yet they involve trade-offs made by the organization… Should the concept of safety culture be broadened to include these kinds of structuring trade-offs? Should they be addressed through a different set of concepts and methods? Sociology of organizations certainly includes part of these questions in its scope.

Yet, a remaining question is: would taking safety into account in all these trade-offs solve all the problems?

3 Is the Solution as Simple as Deciding to Incorporate an Additional Ingredient?

Considering that safety culture in the sense of giving safety sufficient (if not overriding) attention/priority in the trade-offs made by the organization will make all the difference assumes that operating safely is totally in the hands of the organization and that the organization has all the settings to do so. It further assumes that the organization is totally free of its choices and that the problem lies in the lack of internal consistency regarding the attention given to safety. In reality, the situation is complex and the challenges are huge, not only internally but also externally.

Internally, safety models may already vary from one organizational level to the next, one site to another or from one professional culture to another (Carroll, 1998), translating into different approaches to uncertainty management thus to different control mechanisms and eventually requirements (Grote, 2015). An illustration of this variability is the scope of activities that are proceduralized and the philosophy of the rules leaving more or less leeway to those supposed to implement them (Bieder & Bourrier, 2013). Indeed, they reflect the assumptions and beliefs as to what makes operations safe.

In addition, an organization operates in an external environment that not only influences some of its trade-offs but also conveys a mix of perspectives on how safety is to be ensured and ultimately leads to a certain complexity in ensuring safe performance, even if safety is given an appropriate role. A well-known example is the safety regulation(s) the organization has to comply with. Any regulation relies on a safety model in the sense of a set of assumptions or beliefs as to what is needed to ensure safe performance. The regulatory framework thereby imposes a safety model on the organization, constraining some of its choices related to how to ensure safety. This constraint applies whatever the attention given to safety.

Somewhat less obviously, the organization is also constrained by a number of other aspects that include a safety model the organization has very little leeway to negotiate or adapt.

The technological devices used by the organization but not developed in-house also reflect one or more safety model(s). The design of an aircraft or of a control room relies on a number of assumptions as to how safety works (or what is needed to ensure safe operations) that preside over design choices made by the technology designers and manufacturers. These choices are also guided by some assumptions regarding trade-offs between safety and productivity or other business dimensions. A technological device may even reflect several safety models depending on the operational context and conditions. It is the case of degraded modes of automation like in aircraft design for example. The underlying assumptions as to what will ensure that the flight is safe are different, in particular as to the respective roles of the aircraft and the pilots, in normal or degraded modes of automation. The diversity of safety models embodied by the aircraft itself put different requirements on the flight crew depending on the situation. The reality is even more complex.

Indeed, the same technological device may integrate several underlying safety model(s) by combining technologies coming from different sources (e.g. suppliers) that themselves reflect trade-offs between safety and other business objectives. For example, a supplier may favor standardization and thereby adopt a generic "one-size-fits-all" underlying safety model in the design choices fitted for this purpose. In the case of aircraft design, a supplier may provide the same system to several aircraft manufacturers with very limited customization, if any. Eventually, the choice of one technology against another is not neutral in terms of the safety model(s) and trade-offs that it implicitly conveys. One can argue that an airline has the freedom to choose the aircraft and partly the aircraft systems suppliers but anyway, the choice is limited and the leeway remains small.

At this level of detail, the contradictions and inconsistencies appear quite obviously. Nevertheless, there is no point hoping to achieve perfect consistency between all these models. These contradictions are not the sign of an inappropriate way of thinking or of failures of organizations but just a manifestation of the complexity of reality (Morin, 2007). There is no point attempting to reduce them all.

Ultimately, whatever the role given to safety in trade-offs, each organization embeds a mix of safety models that are not necessarily consistent, without even being aware of it. This variety of safety models imposes a set of partly conflicting requirements onto employees at all levels, meaning that "operating safely or safely enough" is not just a matter of safety oriented trade-offs but involves a certain complexity to navigate inconsistencies.

4 Conclusion

Safety culture tends to focus on operations, excluding from its scope trade-offs and decisions that have a structuring and lasting impact on the conditions under which operations take place. Yet, some of these high-level decisions and trade-offs generate inconsistencies "designed" into the organization and its resources that safety culture alone cannot overcome. Could the role given to safety in these higher-level decisions come under an extension of the safety culture concept or is another set of concepts needed from sociology of organizations, management or other disciplines? Whatever the answer, it would most probably contribute to reducing evitable complexity.

However, it is vain to imagine that this complexity, partly due to the coexisting mix of safety models in an organization, can be totally eliminated. An organization operates in an environment that partly constrains its leeway and generates some inconsistencies in the requirements to operate safely or safely enough.

In fact, the situation is even more complex at the macro level. Safety is one aspect among others that needs to be managed by organizations. If the concept of safety culture was introduced to make sure that it was not under-considered/ overlooked compared to others (e.g. productivity, security, financial benefits…), the issue is that of acceptable or globally reasonable trade-offs between the various stakes, rather than protecting a specific one, possibly to the detriment of others, even when this is safety.

References

Amalberti, R. (2015). A way out of the impasse. *Les Tribunes de la sécurité industrielle*, 2015-05. Retrieved from https://www.foncsi.org/en/publications/collections/opinion-pieces/a-way-out-of-the-impasse/tribune-2015-05.

ASN. (2013). *Maîtrise des activités sous-traitées par EDF dans les REP en exploitation*. Retrieved from https://www.asn.fr/L-ASN/Appuis-techniques-de-l-ASN/Les-groupes-permanents-d-experts/Groupe-permanent-d-experts-pour-les-reacteurs-nucleaires-GPR.

Bieder, C., & Bourrier, M. (2013). *Trapping safety into rules, how desirable and avoidable is proceduralization of safety?* Farnham: Ashgate.

Bower, M. (1966). *The will to manage*. New York: McGraw-Hill.

Carroll, J. S. (1998). Organizational learning activities in high-hazard industries: The logics underlying self-analysis. *Journal of Management Studies, 35,* 699–717.

Commission of Inquiry into the Air Ontario Crash at Dryden, Ontario, Canada. (1992).

Grote, G. (2015). Promoting safety by increasing uncertainty—Implications for risk management. *Safety Science, 71,* 71–79.

IAEA. (1991). *INSAG-4, safety culture*. A Report by the International Nuclear Safety Advisory Group. Vienna: International Atomic Energy Agency.

Mearns, K., Kirwan B., & Kennedy, R. (2009). Eighth USA/Europe air traffic management research and development seminar (ATM2009), developing a safety culture measurement toolkit for European ANSPs.

Morin, E. (2007). *Introduction à la pensée complexe*. Paris: Editions du Seuil.

Schein, E. H. (2010). *Organizational culture and leadership* (4th ed.). San Francisco: Wiley.

Chapter 12
Key Issues in Understanding and Improving Safety Culture

Stian Antonsen

Abstract The aim of this chapter is to highlight three key issues that safety researchers and practitioners should consider as part of a cultural approach to the study and improvement of safety. The three issues are: (1) the relationship between integration and differentiation in safety culture research, (2) moving from descriptions of safety cultures to improvement of safety, and (3) the possible downsides of a cultural approach to safety. The chapter argues that a better understanding of boundary processes between groups is vital for a cultural approach to safety improvement and gives a set of general principles for the design of improvement initiatives. Some limits and limitations to the cultural approach to safety are also discussed.

Keywords Safety culture · Integration · Differentiation · Organizational boundaries · Safety improvement

1 Introduction

The concept of safety culture rose to prominence within safety management and safety research around the year 2000. Although the investigation of the Chernobyl accident is ubiquitously cited as being responsible for coining the term, the research origins date further back (e.g. Turner, 1978; Zohar, 1980). The genesis of the concept has been thoroughly described elsewhere (Cox & Flin, 1998; Guldenmund, 2000; Antonsen, 2009a) and will not be repeated here. Rather, in this chapter, I will

S. Antonsen (✉)
SINTEF, Trondheim, Norway
e-mail: Stian.Antonsen@sintef.no

C. Gilbert et al. (eds.), *Safety Cultures, Safety Models*,
SpringerBriefs in Safety Management,
https://doi.org/10.1007/978-3-319-95129-4_12

discuss what I believe to be key issues that should be part of practitioners' and researchers' efforts to adopt a cultural approach to safety improvement.[1]

When scholars and practitioners meet to discuss topics related to safety culture, there are three questions that tend to stand out in the discussions. One is the question of whether an organization can be regarded as having one overarching organizational (safety) culture, or if organizations are better understood by seeing them as consisting of several, sometimes conflicting, subcultures. The other question has to do with the way information and understanding of safety culture(s) can be turned into safety improvement. In this chapter I will reflect on these two questions on the basis of selected literature on organizational culture, as well as my own previous work on the topic. In addition, I will take the opportunity to discuss a third question: what are the possible unintended consequences of adopting a safety culture approach? As researchers and practitioners of risk management, we are trained to look for the different ways in which things may go wrong, and our own efforts of improvement should be no exception to such scrutiny.

2 Integration and Differentiation in Organizational Culture Research

The definition of safety culture has been one of the most widely-debated topics within the safety science research community. Guldenmund's (2000) review of the literature came up with nearly 20 different definitions of safety culture and safety climate, and in the years since his review several more has been proposed. Despite the variety of definitions, there are several recurring themes. One of them is that culture is something that is shared among the members of a social unit, and that it influences what is seen as meaningful ways to behave, communicate and interact. Taking this as a premise, two important questions arise: What are the units of analysis, and what is it that they actually share? These questions have to do with the relationship between integration and differentiation. There is a duality inherent in the concepts of group and culture in that it is both a matter of the internal integration within a group, and the construction of borders with other groups. Thus, understanding integration will most often involve understanding what distinguishes a group from their surroundings and other groups. When dealing with the theoretical and conceptual aspects of culture, this standpoint is fairly uncontroversial. However, for the study and improvement of safety in organizations over a certain size, it becomes more problematic. Is safety culture something that characterizes the organization in its entirety, or are we better off studying safety cultures in plural, thus viewing safety culture as belonging to groups, professions, departments,

[1]The issues raised in this chapter are by no means new to the field of safety science. They have previously been raised and discussed in different ways by authors like Frank Guldenmund, Andrew Hopkins, Knut Haukelid, Carl Rollenhagen and others.

facilities etc. *within* the organizations? The obvious answer is that we should do both: we should both understand the general organizational frameworks that distinguish the organization from other organizations, *and* the differentiation, dynamics and power struggles between the different groups that comprise the organization as a whole. In order to do this, however, we should start by distinguishing between what is shared and what is not shared (Antonsen, 2009b). Safety culture research has been predominately oriented at describing the traits that are shared among the members of an organization. My proposal is to complement this picture by shifting the attention towards the construction of boundaries between groups, and the boundary maintenance[2] that takes place in the interaction between members of groups. In the following sections, I will provide some examples of such boundaries within organizations before turning to a brief discussion of the elements that are likely to be shared across entire organizations.

What can be the possible sources of cultural boundaries within organizations? An organization over a certain size will need to have a horizontal division of labor and a vertical distribution of authority (Mintzberg, 1983) which involve the possibility of cultural differentiation. The division of labor means that people are responsible for different more or less specialized parts of the organization's production processes. Working with the same tasks can be a source of shared identity among the members of a community of practice (Wenger, 1998). The flipside of the coin is that people performing other tasks can be defined as outsiders to this specific community.

The vertical distribution of authority has previously been shown to be the source of cultural differentiation. Schein (1996), often cited as the prime exponent of purely integrative research on organizational culture, describes three different organizational strata which may form different subcultures in organizations: the executive culture, the engineering culture and the operator culture. These groups face different problems and tasks, are likely to have different experiences and perceptions about the organization's activity and may find themselves in conflict and power struggles. Similar descriptions of stratified subcultures have been presented in Johannessen's (2013) study of the Norwegian police and, to some extent, in Crozier's (1964) study of the relationship between workers and management in French industrial organizations.

In addition to, and across these two lines of differentiation, runs a third possible cultural boundary, consisting of the professional identities often shared by people with the same background in terms of education and basic training. Such lines of division are closely related to the previous two, but can cut across organizational strata, departments or even organizations. If you are a doctor, a seafarer, a pilot or a lawyer, you are likely to have some similarities in skills, knowledge or experience with people with the same background, and this can form the basis of dissimilarities.

[2]The term 'boundary maintenance' is borrowed from Barth's (1969) classic discussion of ethnic groups and boundaries.

A last example of cultural boundaries in organization has to do with nationality and ethnicity. As patterns of mobility and migration change, so does the cultural complexity of large organizations. For instance, some studies of Polish construction workers in Norway and England show that there may be differences in the way different groups view the quality of work being performed, the way they interact with people from other groups, and the power distance between managers and superiors where these have different national origins (Wasilkiewicz, Albrechtsen, & Antonsen, 2016; Datta & Brickell, 2009).

The list of possible boundaries described here is by no means exhaustive. It is also important to note that there is no one-to-one relationship between the boundaries described, and cultural differentiation. The existence and nature of cultural differences is an empirical question, not an a priori one. A remaining question, however, concerns the relevance of such cultural boundaries for safety. The answers to this question have much to do with communication and the flow of information in the organization. Turner and Pidgeon launched the concept of "variable disjunction of information" to describe

> a complex situation in which a number of parties handling a problem are unable to obtain precisely the same information about the problem, so that many differing interpretations of the situation exist. (Turner & Pidgeon, 1997, p. 40)

Different people will have access to different information and will also have different frames of reference in interpreting information and situations. Thus, different groups of people will never have precisely the same interpretation of information and situations. This is both a source of requisite variety, and a challenge for organizational communication, interaction and decisions. In any case, knowledge about the various viewpoints of the groups, and the way information is translated when it crosses cultural boundaries, is important knowledge for those aiming to understand and improve the conditions for safety in the organization. For instance, some of the communication across cultural boundaries will regard information about weak signals of danger, improvement measures, documentation of work performance etc. which can prove to be safety-critical in given circumstances.

Some may now wonder what is left of the term 'organization' if it is nothing more than a fragmented collection of subcultures. My point here is not to say that organization-wide integration is impossible. There are a number of factors that may provide the basis of integration across boundaries, five of which I will briefly mention. First, national and ethnic origin is obviously not only a source of differentiation, but also of integration. We are born into our national culture and acquire this culture through primary socialization. An organization that is homogenous in this respect is likely to share the overarching frames of reference of the national culture, e.g. language, fundamental values, general social conventions, etc.

Second, all members of an organization have the same company management. The actions, decisions, communication and behavior of senior managers can be interpreted as expressions of the right way to behave and interact in the organization. The promotion of lower-level managers will also be highly symbolic in indicating which kind of employees are seen as valuable, and how one should act in order to have a successful career.

Third, organizations over a certain size will have an organization structure, a company website, an intranet, a set of operating procedures, and a management system making rules and other formal documentation available for employees in different parts of the organization. These are examples of a formal context that the members of an organization usually have to relate to in one way or another.

Fourth, most organizations have a set of espoused values describing how key actors would like the organization to be perceived, both by its employees and its surroundings. When these are known among the members of the organization (which is not always the case) they form common points of reference for what is stated as the desired form of behavior.

Fifth, some organizations have experienced crisis, disaster or other key events that stands out in the organization's recent history. The way these are being interpreted by the different groups in the organization may not be characterized by neither clarity nor consistency, but they will still form a point of reference that cuts across the lines of differentiation in the organization.

What is likely to be shared across the subcultures of an organization is thus related to national culture, a general formal context, and key events that have occurred in the organization's recent history. What is not shared are the micro-level experiences that create, recreate and change the informal aspects of organizing. The comments above illustrate that discussions about sharing and differentiation are closely intertwined. The very existence of differences between groups presupposes a level of integration within each of the groups. The important point in this respect is that an empirical question is raised not only by the content of integration, but also by the level of that integration.

3 From Description to Improvement: How Do We Move from Diagnosis to Treatment?

One of the questions I have frequently been asked by practitioners is "*what is the best way to improve our safety culture?*" My answer is always the same: "*There is no one best way to improve your safety culture*". Selecting the strategy of improvement depends on the problem that needs to be solved, and the kind of improvement practices that are likely to resonate with the people owning the problem. Expecting there to be one, proven approach that works irrespective of company, context, history and problem is like expecting your doctor to be able to prescribe a medicine that is guaranteed to improve your health and wellbeing, irrespective of your symptoms, history of illness and underlying health condition. While I have no doubt that my answer is correct, I cannot help but feel that I should be able to come up with a list of examples of the way different approaches have contributed to solving different problems in different contexts, and a more scientific discussion of *why* efforts to improve safety have succeeded or failed.

This points to what I believe is a gap in safety culture research: the body of research articles describing safety culture by far outnumbers the articles reporting the results of efforts to turn descriptions of safety culture into safety improvements. There are probably many reasons for this. One is that improvement projects are often run by consultants, not researchers. Hence, publishing the projects' results are not a priority task. Another reason is that safety improvement is hard to measure, particularly when it comes to major accident risk, and thus it is hard to convince the editors of scientific journals that the results from improvement projects are worth publishing. This should be seen as a joint challenge to high-risk industries and the research community. There is a need to empirically document improvement efforts with a greater level of scientific rigor and 'objectivity' in order to create a better repertoire of possible safety improvement strategies. There is also a need to connect the field of safety improvement to the general literature on organizational improvement. In my previous publications, I have tried to extract some lessons from the general literature on organizational development that, in my view, can form a platform for improving safety through a cultural approach. The result was a list of ten general principles:

1. Organizational transformation is self-transformation. If we accept the premise that cultures are created from the interaction between people, the various groups and professional communities need to be involved in dialogue to define both problems and solutions.
2. Goals should be moderate and relate to everyday realities. Unless improvement measures can be related to the daily tasks and reality of the problem owners, it will not have lasting effects.
3. Change must be viewed as a long-term project. While the boxes and arrows of an organization chart can be moved in a matter of minutes, changing the way people work requires years of persistence.
4. The goal should not be organization-wide consensus, but creating a common language and understanding *between* groups. Differentiation between groups can be a vital resource for safety. Multiple perspectives are a source of requisite variety that can increase the chance of weak danger signals being detected somewhere in the organization.
5. Combine 'push' and 'pull'. Top-down change efforts will fail unless there is a motivation for change at the sharp end of the organization.
6. Management initiates and contributes, but the shop floor must be involved in continuous dialogue. Personnel on the ground have expert knowledge on hazards, work processes and situational demands. This must be acknowledged, respected and utilized.
7. Be sensitive to organizational symbolism. For instance, organizational stories are powerful conveyors of culture and can emphasized to illustrate the problems to be solved, or the early wins in the change process.

8. Gaps between frontstage visions and backstage priorities will derail change processes. If people stop trusting the intentions and truthfulness of managers or coworkers, the change process can have very unpredictable outcomes and negative consequences.
9. Be sensitive to local sense making. Customize change efforts to the different groups that need to be involved to achieve change.
10. Consider the need for change and the realism of objectives. The 'change-or-die' mantra is overrated, particularly when it comes to safety. If you are not sure that you have an organizational problem related to safety, don't start fixing it.

Although these principles are of a rather general nature, I would still like to express a few reservations about their application. Much of the research behind the principles are performed in Scandinavian organizations. The working life tradition in Scandinavia emphasizes worker participation, empowerment and a high level of job security. This is a framework condition that no doubt exerts influence on which improvement strategies are likely to succeed and fail. This means that we should be wary of transporting improvement measures across contexts without consideration of differences in culture and framework conditions.

A final note on cultural change is needed with reference to learning from major accidents and disasters. An important part of safety improvement, as in all other efforts to improve organizations, is to 'unfreeze' the existing structures (Lewin, 1947) before changes can be made. Major accidents provide organizations with the strongest possible motivation to engage in critical reflection on matters related to safety. This introduces the post-accident phase as a key window of opportunity to influence basic safety assumptions. An accident or a disaster leaves the organizations involved shaken to their very foundations. Although there is not necessarily agreement on diagnosis or treatment, there is still an urgent impression that maintaining the status quo is not an option. The shock of the disaster, the investigation reports, media coverage and organizational stories can constitute boundary objects that function as a common point of reference for communication and decision-making that cuts across internal organizational heterogeneity and other organizations in the same industry. For instance, the capsizing of the Alexander Kielland oil rig is still a major reference point for safety in the Norwegian petroleum industry, and few accidents have gained more worldwide attention than that of Deepwater Horizon. This means that major accidents or disasters constitute a source of common experience that may be shared across contexts, as described above. Safety cultures will be strongly influenced by the adverse events that the group(s) have either experienced themselves or can relate to in terms of industry, work situation or profession. Part of a cultural approach to safety improvement involves moving people's horizons of understanding of what can go wrong, and this horizon is strongly influenced by the stories and experiences of major accidents.

4 The Downside of Cultural Explanations for Safety

The cultural approach to safety was the focus of much optimism around 2000, when several companies launched massive safety campaigns. While this optimism is now somewhat downgraded, it is still fruitful to consider the limits to the safety culture approach from time to time.

There is an English saying stating that "*no good deed goes unpunished*" which essentially means that however good the intentions of an action may be, there will always be some unintended consequence that can cause benevolent actions to backfire. This is also true for the concept of safety culture. One downside lies in the use of culture as an explanatory variable for accidents. Saying that an organization's culture has contributed to creating an accident implies that something was wrong with that particular organization. Attributing causality to the unique characteristics of particular organizations makes it easy to conclude that "it could not happen here", and thus close the door toward learning from other organizations' accidents. Consider the very origins of the concept of safety culture, the Chernobyl accident. The accident shook the nuclear industry to its foundations throughout the world. Without reverting to conspiracy theories, attributing the causes of the accident to the cultural traits of the Soviet system (as well as the particular reactor design) can be a convenient way of reestablishing the belief that the errors that could occur in a Soviet organization were unthinkable in the Western part of the industry. A similar point has been made by Bye, Rosness, and Røyrvik (2016) in their analysis of the use of the term 'HSE culture' in investigation of incidents in the Norwegian petroleum industry. They found that the term 'poor HSE culture' led to premature closure in the search for an accident's causes as some uses of the term entailed little other than simplistic explanations of rule violations. Consequently, the use of safety culture as an explanatory variable in accidents investigations should be used with some caution if the aim is to facilitate learning in other organizations.

Another possible downside of focusing on safety culture lies in the relationship between the 'hard' and 'soft' aspects of safety. Previous research has discussed whether the focus on safety culture can be an excuse for not investing in new technology and developing technological design. This is an important question. Measures directed at controlling behavior are usually cheaper than changing production technology. However, focusing on behavior involves measures that compensate for the existence of a hazard. It is less suitable for removing the source of danger causing the problems in the first place. Focusing resources on safety culture can thus be a source of false security if it replaces the continuous search for safer technology.

5 Conclusion

The initial enthusiasm that surrounded the concept of safety culture around the year 2000 has now (fortunately) waned and has been replaced by a more realistic and mature approach in terms of theoretical grounding and methodical approach. I have argued that a key to taking the next step is to better take into account the boundary processes between groups. The point is that if we do not look for such boundaries and boundary processes, we are likely to overlook them, and if we overlook them we miss a great deal of the dynamics that constitute organizational life. As safety is usually a composite 'product' of the efforts of several groups, the dynamics between integration and differentiation needs to be addressed.

An important next step in safety culture research is to study safety interventions. Creating new knowledge about the improvement of safety by means of a cultural approach will require close collaboration between researchers and problem-owners. There is a need for more long-term research collaboration between academia and industry to ensure the realism and rigor needed to design and document high-quality development processes.

References

Antonsen, S. (2009a). *Safety culture: Theory, method and improvement*. Farnham: Ashgate.

Antonsen, S. (2009b). Safety culture and the issue of power. *Safety Science, 47*(2), 183–191.

Barth, F. (1969). Introduction. In F. Barth (Ed.), *Ethnic groups and boundaries: The social organization of culture difference*. Oslo: Universitetsforlaget.

Bye, R. J., Rosness, R., & Røyrvik, J. O. D. (2016). 'Culture' as a tool and stumbling block for learning: The function of 'culture' in communications from regulatory authorities in the Norwegian petroleum sector. *Safety Science, 81,* 68–80. https://doi.org/10.1016/j.ssci.2015.02.015.

Cox, S., & Flin, R. (1998). Safety culture: Philosopher's stone or man of straw? *Work Stress, 12,* 189–201.

Crozier, M. (1964). *The bureaucratic phenomenon*. Chicago: University of Chicago Press.

Datta, A., & Brickell, K. (2009). "We have a little bit more finesse, as a nation": Constructing the Polish worker in London's building sites. *Antipode, 41*(3), 439–464. https://doi.org/10.1111/j.1467-8330.2009.00682.x.

Guldenmund, F. W. (2000). The nature of safety culture: A review of theory and research. *Safety Science, 34,* 215–257.

Johannessen, S. (2013). *Politikultur: Identitet, makt og forandring i politiet*. Trondheim: Akademika.

Lewin, K. (1947). Frontiers in group dynamics. In D. Cartwright (Ed.), *Field theory in social science*. London: Social Science Paperbacks.

Mintzberg, H. (1983). *Structure in fives: Designing effective organizations*. Englewood, New Jersey: Prentice Hall.

Schein, E. H. (1996). The three cultures of management: Implications for organizational learning. *Sloan Management Review, 38*(1), 9–20.

Turner, B. (1978). *Man-made disasters*. London: Wykenham Science Press.

Turner, B. A., & Pidgeon, N. F. (1997). *Man-made disasters* (2nd ed.). Oxford: Butterworth Heinemann.

Wasilkiewicz, K., Albrechtsen, E., & Antonsen, S. (2016). Occupational safety in a globalised construction industry: A study on Polish workers in Norway. *Policy and Practice in Health and Safety, 14*(2), 128–143.

Wenger, E. (1998). *Communities of practice: Learning, meaning, and identity*. Cambridge: Cambridge University Press.

Zohar, D. (1980). Safety climate in industrial organizations: Theoretical and applied implications. *Journal of Applied Psychology, 65*(1), 96–102.

Chapter 13
Safety Cultures in the Safety Management Landscape

Jean Pariès

Abstract As an emerging scientific concept, the notion of "safety culture" presents obvious difficulties. But this does not preclude that it can be quite useful for the management of safety. However, the usual understanding of the concept lacks a reference to an explicit safety paradigm. It describes *organizational* features that are expected to foster safety, but does not explicitly mention the underlying assumptions about the safety strategy expected to make the *system* safe. Yet, there is no one single strategy to make a system safe. Even within a given organization, there must be a variety of strategies, with a different balance between predetermination and adaptation, and different levels of control on front line operators. Each of these safety management modes will inevitably generate the corresponding "safety culture". The underlying safety management mode behind the current safety culture vision is a non-punitive version of a normative and hierarchical safety management mode. However, evolving toward this mode does not necessarily mean that safety culture is becoming more mature. Recent catastrophic accidents have illustrated the increasing vulnerability of our systems to the unexpected, and illustrated the need for a refined safety paradigm.

Keywords Safety culture · Safety paradigm · Safety management modes
Predetermination · Adaptation

1 A Brief Historical Perspective on Culture and Safety

The impact of culture on the performance of organizations has become a growing concern in western industries with the globalization of companies. This has led to the development of a whole line of research, particularly well illustrated by the seminal cross-cultural work of Hofstede (1980, 1991). Defining a culture as

---

J. Pariès (✉)
Dédale SAS, Paris, France
e-mail: jparies@dedale.net

C. Gilbert et al. (eds.), *Safety Cultures, Safety Models*,
SpringerBriefs in Safety Management,
https://doi.org/10.1007/978-3-319-95129-4_13

the collective programming of the mind which distinguishes the members of one group from another,

Hofstede tried to identify the impact of national (ethnographic) cultures on organizational (corporate) cultures. His conclusions laid the foundation for a considerable body of work that has examined the role of national cultures *in relation to safety*, particularly in aviation. A Boeing study (Weener & Russel, 1993) showed that for the years 1959–1992, the proportion of accidents in which the crew was considered a causal factor varied in a ratio of about one to five with respect to the region of origin of the airline. Merritt (1993, 1996) replicated Hofstede's work to explore cross-cultural similarities and differences with respect to attitudes toward flight management and the link to safe operations. Her findings paralleled Hofstede's in revealing significant differences in attitudes toward authority and the extent to which people preferred to make decisions individually or via consensus. Helmreich and Merritt (1998) explicitly searched for correlations between national particulars with respect to Hofstede's cultural dimensions, and the accident rates of airlines. They found that (only) two of these dimensions were correlated with safety performance: power distance and collectivism/individualism.[1]

Many then concluded that differences in national culture caused pilots from Asia, Africa, or South America to be less safe than those from the USA or Europe. However, the same research also included outcomes in dissonance with this vision. The 1993 Boeing study also showed that non-western operators did not suffer a higher accident rate than western ones when compared on the same routes. Helmreich and Merritt (1998) clearly rejected the link between national culture and accident rates:

Some authors have correlated national culture with accident rates and concluded that pilots in certain countries are safer than others. We take umbrage with the simplicity of this statement. The resources allocated to the aviation infrastructure vary widely around the globe. {...} Accident rates are a function of the entire aviation environment, including government regulation and oversight, and the allocation of resources for infrastructure and support, not just pilot proficiency (p. 104–5).

To test the relationship between accident rates and infrastructure, Hutchins, Holder, and Pérez (2002) performed a correlational analysis across major regions of the world on common measures of infrastructure quality and a measure of flight safety. They show that flight safety is correlated at the 0.97 level with daily caloric intake.

If a nation does not have the wealth required to create and distribute food, it is unlikely to be able to invest in modern radar systems, ground-based navigation and approach aids, runway lighting, weather prediction services, or the myriad other institutions on which safe civil aviation operations depend.

[1]Uncorrelated dimensions are uncertainty avoidance and Masculinity/Femininity.

As noted by the authors themselves, this does not mean that culture plays no role in the organization of (safety-related) behavior on the flight deck. But it means that culture is only one of a large number of interacting behavior drivers, so that its relative effects on behavior are unknown and may remain so.

Furthermore, there is a chicken-and-egg issue between culture, infrastructure, and behavior. Culture is commonly seen as a set of shared behavioral attractors (values, beliefs, attitudes) literally written in people's minds. In this vision, *culture shapes behavior*. But conversely, people also behave in certain ways because they make sense of their situations, define their own goals to serve their interests, and act accordingly. When environments, goals and interests are similar, behaviors tend to be similar. They reinforce each other through imitation, and crystallize into binding stereotypes that become values and attitudes. In this vision, *environments and behaviors generate culture*. So culture shapes behaviors which mold infrastructure that influence behaviors that crystalize into culture. They are linked as the ingredients of an autopoietic system (Maturana & Varela, 1980). A forest does not last as a forest merely because trees reproduce themselves. A forest permanently regenerates, through the transformation, the destruction and the interaction of its components, the network of components' production processes and the environmental conditions needed for its regeneration.

The focus has later shifted from *national* to *organizational* culture. The idea that organizational or corporate culture—defined as the reflection of shared behaviors, beliefs, attitudes and values regarding organizational goals, functions and procedures (Furnham & Gunter, 1993)—can by itself shape safety behavior, hence safety performance within an organization, is indeed an attractive assumption. However, this definition suffers from the same fundamental ambiguities as ethnographic culture. First, it does not solve the circularity between culture, behavior, and environments, and we find definitions that simply include what people think (beliefs, attitudes and values), and others that also include how people act (behaviors). A reference in the latter category, Schein (1990, 1992) suggested a three-layered model including (i) core underlying assumptions, (ii) espoused beliefs and values, and (iii) behaviors and artefacts. Second, the notion of corporate culture postulates by definition some autonomy from national cultures, but does not define the extent of this independence. Through the reference to *shared* beliefs and values, it also assumes a certain level of internal cultural consistency within a given organization, but it is not clear how this postulated 'cultural color' treats obvious internal sub-cultures, i.e. the differences between trades and groups within the same organization. Last but not least, the assumption about its impact on safety is unproven, although plausible.

2 The Birth of "Safety Culture": Not Rocket Science but a Useful Concept

The reference to the term "safety culture" by AIEA in the aftermath of the 1986 Chernobyl disaster (INSAG-1, 1986; INSAG-4, 1991; INSAG-7, 1992) can be seen as a further attempt to clarify the link between culture and safety. Safety culture was defined as

> that assembly of characteristics and attitudes in organizations and individuals which establishes that, as an overriding priority, nuclear plant safety issues receive the attention warranted by their significance.

The number of available definitions in the academic and corporate literature shows both the success of the concept and its ambiguities. The common point of these definitions is that safety culture is *the sub-set of corporate culture* that influences safety (this establishes a link to safety, at least in theory). As with corporate culture, these definitions mainly differ according to whether or not they include behavioral patterns. The AIEA definition belongs to the "non-inclusion" family, and mainly reflects the concern and commitment to safety (usually called *safety climate*). The other family of definitions includes behavioral patterns and reflects both commitment and competence to manage safety. The definition given by the UK Health and Safety Commission is a prominent representative of this family:

> (Safety culture is) the product of individual and group values, attitudes, competencies, and patterns of behavior that determine the commitment to, and the style and proficiency of, an organisation's health & safety programmes. (HSC, 1993)

The difficulty with the former family is that (safety) behaviors do not result from cultural influence only. The difficulty with the latter is that cultural influence does not determine (safety) behaviors in a straightforward, deterministic way. Managers tend to prefer the former, because it explicitly refers to what they seek to influence: behaviors.

It is difficult to manage something if it is not assessable, but the assessment of safety culture poses further challenges. As noted by Hutchins et al. (2002)

> {…} in order to assess the value of a culture to {…} safety, one would have to cross all available cultural behavior patterns with all conceivable {…} circumstances. In every case, one would have to measure or predict the desirability of the outcome produced by that cultural trait in that particular operational circumstance. Constructing such a matrix is clearly impossible.

Instead, the use is made of surveys measuring attitudes or self-reported behaviors against attitudes or behaviors that have been considered by safety experts as leading to desirable (or undesirable) safety outcomes. As the industry lacks the observational data to match attitudes with real behaviors in operational contexts, and even more to match behaviors with safety outcomes, this kind of assessment grid is merely a mirror of the questionnaire designers' current vision of the influence of attitudes and behaviors on safety *in their own culture*. In other words, safety culture can hardly be regarded as a scientific concept, and when it comes to assessing it, safety culture is more or less implicitly defined as "what is measured by my survey".

3 Safety Culture and Safety Paradigms

It does not follow that 'safety culture' is an irrelevant or unworkable concept for safety management. Safety culture assessment surveys provide an interpretation of behavior-related safety issues (Cooper, 2000). Taken with due precaution considering their conceptual ambiguity, these "quantified" pictures can be an effective starting point for discussing behavioral dimensions of safety management within an organization. Indeed, the assessment process can actually start with the efforts to interpret the outcomes of the survey. The apparent "objectivity" of the survey results, discussed during interviews and focus groups, helps the organization's members to step back and look at themselves as if in a mirror. Even if the mirror is highly distorted, it triggers the perception of an image—or a caricature—of the organization. The collective sense-making process about this image can bring about the questioning needed and the potential triggers for a change.

In my experience, a key issue is that *these pictures do not offer much reliable and objective meaning* per se: the answers to many typical survey questions can lead to several different, plausible and often contradictory, interpretations. For example, the following assertions[2] are extracted from a Eurocontrol safety culture questionnaire: "*Sometimes you have to bend the rules to cope with the traffic*"; "*Balancing safety against other requirements is a challenge—I am pulled between safety and providing a good service*". Would disagreement mean adherence to rules and giving a high priority to safety, hence a "good" safety culture, or would it mainly represent a high degree of jargon and an unrealistic perception of actual safety challenges? The arbitration between these alternative interpretations requires additional data about real work, behaviors and infrastructures, at the scale of a work group. It also requires an interpretation grid—a gauge—to make sense of the answers for the different trades. The variability across trades within the same organization may well be much higher than the inter-organizations variability within the same trade. Hence I am skeptical about the meaning of cultural benchmarks based on the same questionnaire across different organizations.

Safety culture questionnaires necessarily convey underlying and implicit assumptions about what enables an organization to stay in control of its safety risks. Usual assumptions include the commitment of managers and staff to safety, clear and strictly-obeyed rules and procedures, open and participatory leadership, good synergy within teams, open communication between colleagues—and through

[2]Respondents are requested to express their agreement/disagreement on a five-degree Likert scale.

hierarchical layers, transparency about failures and incidents. All these assumptions reflect the opinions of safety experts about what attitudes and behaviors lead to desirable safety outcomes. They appear rational and to be common sense. They appeal to managers because they are manageable, and in line with the established order: management is responsible for designing and defining the "right" behaviors, leading and "walking the talk"; front line operators' are responsible for complying with the prescriptions and reporting difficulties and failures. They refer to safety indicators based on measurable and controllable events frequencies. They have made the fortune of DuPont and a few others.

However they lack an explicit reference to a clear safety paradigm. They describe *organizational* (managerial, cultural) features that are expected to foster safety, but they do not explicitly mention the underlying beliefs about what makes a *system* safe. They address the syntax of safety management rather than its semantics. As a consequence these assumptions are difficult to "falsify"—in Kuhn's (1996) terminology—by factual evidence. Hence they tend to become unquestionable dogmas. The famous assertion of a constant ratio between unsafe behavior, minor injuries, and fatal accidents (Heinrich, 1931; Bird & Germain, 1985) is a first example. This belief has been used worldwide throughout the industry for decades to prevent severe accidents through chasing daily noncompliance and minor incidents. Yet, it has been refuted by many researchers (Hopkins, 1994, 2005; Hovden, Abrechtsen, & Herrera, 2010) and characterized as an "urban myth" by Hale (2000). A recent study conducted by BST & Mercer (Krause, 2012; Martin, 2013) on occupational accidents in seven global companies (ExxonMobil, Potash Corp, Shell, BHP Billiton, Cargill, Archer Daniels Midland Company and Maersk) clearly shows a de-correlation between the evolution of the fatal and non-fatal accident rates over a given period. Barnett and Wang (1998) reached similar conclusions about the link between airlines incident rates and the mortality risk of passenger air travel over a decade (1987–1996) in US flight operations. In plain language, it means that safety strategies about severe accidents based on the Bird pyramid are at least partially flawed and inefficient, whatever their intuitive attractiveness and commercial success.

A second example is the moral posture embedded in most safety culture assessments concerning errors and violations. The acceptance that "errors are inevitable" is seen as a positive safety culture trait, while the acceptance for intentional deviations is seen as very negative. However, the respective contributions of errors and violations to the safety risk, when quantified,[3] did not necessarily support the above judgement. Within the Line Operations Safety Audit (LOSA) program in aviation (Helmreich, Klinect, & Wilhem, 2000), specifically trained senior pilots observe from the jump seat anonymous crews managing safety risks during real flights, and assess the risk generated by external and internal threats, actions, and inactions, in the various situations faced. Not surprisingly, deviations could be observed on 68% of flights and the most frequent were violations.

[3]Which has been rarely done yet.

More interesting is the assessment of the associated risk: only 2% of the violations were classified as consequential, in contrast with 69% for proficiency-related errors. As this was not in line with the dominant beliefs in aviation—violations must matter—further analyses were conducted to demonstrate that

> those who violate place a flight at greater risk. {…}. We found that crews with a violation are almost twice as likely to commit one of the other four types of error and that the other errors are nearly twice as likely to be consequential.

Interestingly, the reverse hypothesis that violations could be a consequence of errors (e.g. attempts to mitigate errors) has not been envisaged….

4 Safety Management Modes

There is no one single strategy to make a system safe, which would work regardless of the system, its design, its business model and its environment (Amalberti & Vincent, 2014; Grote, 2012). The Bird triangle may work in some contexts while not in others. A total compliance with procedures may be an absolute safety condition in some contexts, and a threat in others. So there is a need for a generic grid of *safety management strategies,* which would allow for, and make sense of, different weights of the syntactic dimensions of safety culture (compliance, transparency, autonomy, accountability…). Safety management is inseparable from uncertainty management (Wildawsky, 1988; Westrum, 2006; Grote, 2007). It is also totally dependent on the way an organization generates, through its different layers and through its design, the behaviors that maintain the system in a safe state. The observed diversity of safety management strategies can hence be seen as the result of a combination of two key features: (i) the nature and level of predetermination in the management of uncertainty, and (ii) the nature and level of centralized control on front line operators.

These two features are usually considered interdependent—hence they are merged—in safety management theories. Anticipation and predetermination are considered to imply a centralized and hierarchical bureaucracy with a high level of control over operators. Conversely, resilience and responsiveness would imply a flexible, self-organizing organization. Amalberti (2001, 2013) describes a linear continuum of safety management modes ranging from "resilient" systems to highly normalized "ultra safe systems". Journé (2001) suggests that the articulation of uncertainty management and organizational features leads to the definition of two "safety management systems": a *mechanist* model, based on rational anticipation and bureaucratic organizational control, and an *organic* model, based on resilience, a decentralized organization, and the self-organizing capacities of autonomous teams. Grote (2014) proposes a more sophisticated correspondence grid between uncertainty management strategies (reducing/absorbing/creating uncertainty) and organizational control modes on operators. However, her approach still seems to be based on interdependent associations between uncertainty management modes and organization features. My contention is that these two features are much less interdependent than usually assumed. Instead, they define two independent dimensions, hence a two-dimensional space that can be summarized with four main combinations, defining *four basic safety management modes* illustrated by Fig. 1.

In quadrant 1, a combination of high predetermination and strong organizational control enables a centralized risk management. The system is designed to be safe, and the strategy is to stay within its designed-to-be-safe envelope, which is continuously refined and expanded through in-service experience feedback and quality improvement loops. Predetermination of responses, planning, compliance with norms and standards, as well as hierarchical control, reduce many of the existing variability dimensions. Front line operators are highly standardized through selection and training, and are interchangeable. The power and responsibility for safety belong to the central organization.

In quadrant 2, in contrast, a combination of low predetermination and low organizational control leaves each frontline operator or team with the responsibility for managing the trade-offs between safety and performance. These are generally open systems, operating in an environment characterized by a high level of unpredictability. Their responses cannot be easily predetermined or standardized. Norms and regulations are only partially effective for safety. They must be complemented by strong adaptation expertise. Safety mainly emerges from adaptive processes and self-organization. The power and responsibility for safety belong to front line managers and operators.

In quadrant 3, a strong hierarchical organizational control is exerted over front line operators (by means of authority, intensive training, strict compliance with rules and procedures…). But operations need to be highly adaptive, because the anticipation possibilities are low, due to a high degree of uncertainty in the situations faced. In these systems, a highly effective, "maestro" type, operational hierarchy evaluates situations, makes decisions and adapts the responses, commanding highly trained and disciplined front line actors, acting in a tightly coordinated and standardized way. The power and responsibility for safety mainly belong to the operational commanders.

Finally, in quadrant 4, a combination of high predetermination and low organizational control allows the system to operate in an environment with strong constraints of operational conformity, while handling high variability in the details of operational situations. The decision power is delegated to local structures directly coupled to real time activity. But their overall behavior is to a large extent predetermined. These are highly cooperative systems, in which global behavior emerges from networking the activity of multiple autonomous cells. Front line players have similar skills, they follow rules and procedures, but their real time behavior is controlled by a strong team culture. The power and responsibility for safety mainly belong to operational teams.

Figure 1 also gives a few examples of potential representative domains of activity for each of these safety management modes. However, it is important to note that these examples must be taken as a caricature of a much more complex reality.

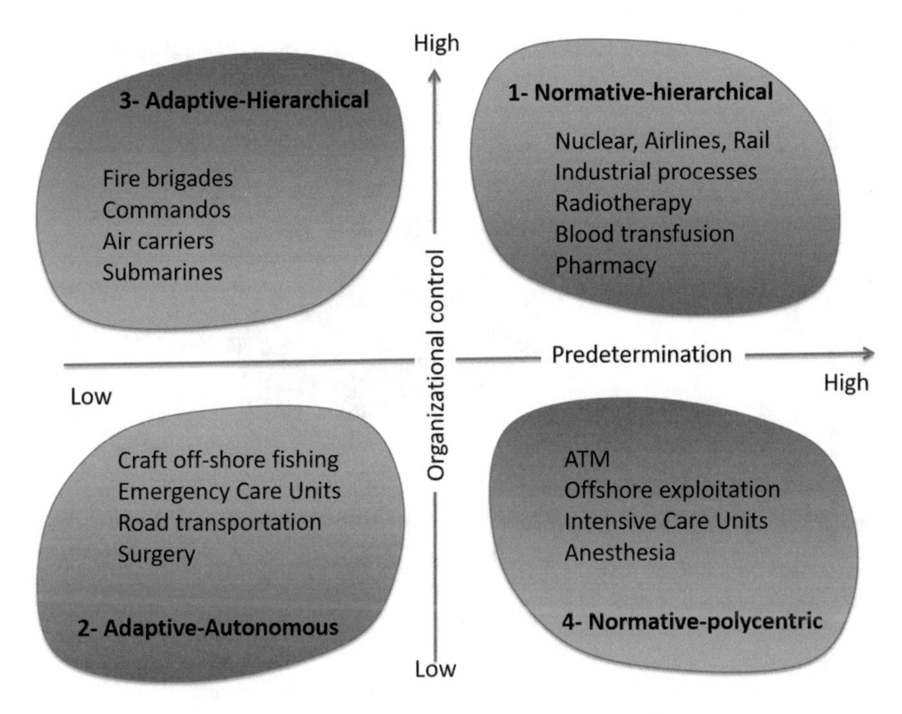

Fig. 1 Basic safety management modes

They only refer to the dominant safety philosophies in each area. In fact, the different components or business units of a large organization would spread across several modes as illustrated by Fig. 2.

5 Safety Culture and Safety Management Modes

A variety of safety management modes should be regarded as both normal and desirable within a large organization. Indeed the balance between predetermination and adaptation should be coherent with the actual level of endogenous and exogenous uncertainty. And the organizational control on individual safety behaviors must be in coherence with the social realities of the organization: the management of safety cannot be based on a type of social relationship significantly diverging from the overall management style. But the levels of uncertainty may be very different from one activity to another, even for apparently similar activities. For example, in Air Traffic Control Services, aerodrome control must handle much more uncertainty than en route control, because it needs to accommodate general aviation and private pilots. Similarly, the power distribution between trades or across hierarchical levels, usually resulting from a long confrontational history, may be very different within various components of an organization. Hence the senior

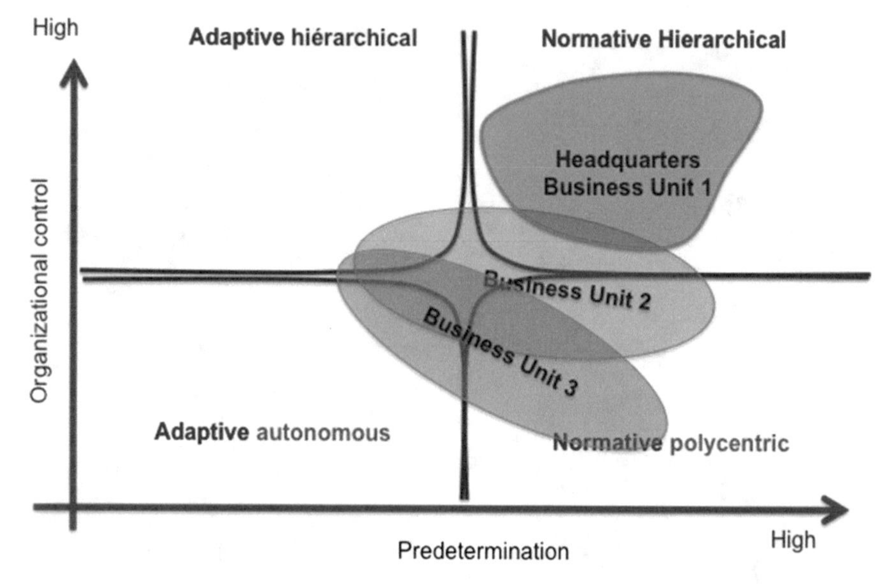

Fig. 2 Illustrative representation of a given organization

management should recognize the need for the corresponding diversity, and explicitly foster it rather than try to reduce it.

But then, should there be a corresponding diversity within the organizations' safety culture? What is the relationship between safety culture and safety management modes? It is not a one-to-one relationship. In the long term, each safety management mode tends to generate its own sub-culture. However, reversely, organizational cultures tend to persist for a long time, and may prevent the development of safety culture traits consistent with an emerging safety management mode. As can be readily observed during mergers, cultural misalignments can persist for years within the resulting company. Similar cultural misalignments can be encountered within a given company, between central and regional structures, corporate level and business units, trade or front line practices and managerial expectations. They manifest themselves through latent conflicts such as this typical example: managers invoke safety to try and reinforce their authority on their staff. Symmetrically, their staff resist procedures and Unions defend indefensible unsafe behaviors, as a resistance weapon against authority.

Beyond this, a significant part of the underlying safety management values and rationalization is imported from national beliefs and demands, as well as from international standards. The currently dominant vision of the "ideal" safety management mode in these standards is a soft (non-punitive and very Anglo-Saxon) version of the normative-hierarchical mode. However, evolving toward this mode does not necessarily mean a safer system or a march toward a higher safety culture maturity. Despite its undisputable contribution to historical safety progresses, the "total predetermination" strategy has also shown limitations. Recent catastrophic accidents (AF447, Fukushima, Deep Water Horizon) have illustrated the increasing vulnerability of large sociotechnical systems to the unexpected and the need for a refined safety paradigm. However, two powerful socio-cultural mechanisms continue to feed the trend towards more norms and compliance. The first is the dominant "positivist" culture of designers and managers, who perceive safety as the result of a deterministic, top-down, command-and-control process. The second is the increasing pressure of legal liability on the different players, including policy makers, requesting everyone to demonstrate that "risks are under total control". So people may develop a vision of safety based on 'nit-picker' compliance, not so much because they rely on objective safety performance outcomes in their activity domain, but rather because they seek to minimize their liability.

In brief, Safety Culture inevitably and inextricably incorporates dimensions of organizational and national cultures that do not directly emerge from the reality and rationality of safety management modes, and can even be in conflict with them. Coherence is desirable, but conflictuality is not necessarily something bad. As discussed earlier, the culture-behavior-performance relationship is not a linear one. As with the bow, the strings and the violin, it is rather a resonance, whose equilibrium point cannot be foreseen. Tension and friction are needed to play music. And even a dose of bluff: one must sometimes 'preach the false for the true knowledge', demand total obedience to get intelligent compliance, value errors to build confidence. Hence a safety policy should not be based only on beliefs, but also on facts. In the semantics of safety cultures, evidence-based safety management should frame, if not replace, assumptions and dogmas. This in turn implies that, in each activity domain, 'work as really done' is properly assessed and relevant metrics are developed—and implemented—to measure things such as the level of uncertainty, the contribution of non-compliance to safety risk, or the statistical correlation between the frequency of small deviations and the likelihood of disaster.

What is at stake behind the notion of uncertainty is not only its extension, but also the very nature of uncertainty. In simple systems the impact of events is well known, so decisions only depend on the probability of occurrence, which is usually well expressed by Gaussian distributions. In complex systems, the probabilistic structure of randomness may be unknown or misjudged (the distribution tail may be much thicker than expected), and there is an additional layer of uncertainty concerning the magnitude of the events. In this case the risk associated with the unexpected may be far greater than the known risk. Focusing on anticipations and failing to "manage the unexpected" thus echoes the story of the drunk looking for lost keys under the lamppost "because the light is much better here". Ironically, safety management—which is about managing uncertainty—may be eventually impaired by the illusory byproduct of its success: a rising culture of certainty.

6 Conclusion

The concept of 'safety culture' may not be a scientific one, but this does not preclude it from being quite useful for the management of safety. Safety culture assessment surveys provide an interpretation of behavior-related safety issues. Taken with due precaution this can be an effective starting point to discuss behavioral dimensions of safety management within an organization, in order to initiate a change. However, even from this pragmatic perspective, the usual acceptation currently describes organizational features expected to foster safety, but does not explicitly mention the *underlying beliefs* about the safety strategy to keep the system safe. Yet, there is no one single strategy to make a system safe. Depending on endogenous and exogenous uncertainty, there must be a variety of strategies providing different trade-offs between predetermination and adaptation, as well as different ways of exerting control on the behavior of front line operators. Coherent combinations define safety management modes.

But there is no one-to-one matching between safety management modes and safety cultures. A safety culture inevitably incorporates 'local' as well as organizational and national dimensions that do not directly emerge from the rationality of safety management modes, and can even be in conflict with them. This is not necessarily a problem, but an arbitration judge is needed, and factual evidence is best. In the semantics of safety cultures, evidence-based safety management should take priority over assumptions. This in turn implies that in each area of activity, 'work as done' is properly understood and relevant metrics are developed and implemented in order to measure the level and nature of uncertainty, i.e. the correlation between small deviations and the likelihood of disaster.

References

Amalberti, R. (2001). The paradoxes of almost totally safe transportation systems. *Safety Science, 37,* 109–120.

Amalberti, R. (2013). *Navigating safety: Necessary compromises and trade-offs—Theory and practice*, Springer Briefs in Applied Sciences and Technology Heidenberg: Springer.

Amalberti, R., & Vincent, C. (2014). A continuum of safety models. *Risk Dialogue Magazine*, http://institute.swissre.com/research/risk_dialogue/magazine/Safety_management/A_continuum_of_safety_models.html.

Barnett, A., & Wang, A. (1998). *Airline safety: The recent record.* Massachusetts Institute of Technology. National Center of Excellence in Aviation Operations Research.

Bird, F. E. J., & Germain, G. L. (1985). *Practical loss control leadership*. Loganville, GA: International Loss Control Institute Inc.

Cooper, M. D. (2000). Towards a model of safety culture. *Safety Science, 36,* 111–136.

Eurocontrol Safety Culture in ATM—An overview. (2008). http://www.eurocontrol.int/articles/safety-culture.

Furnham, A., & Gunter, B. (1993). *Corporate assessment*. London: Routledge.

Grote, G. (2007). Understanding and assessing safety culture through the lens of organizational management of uncertainty. *Safety Science, 45,* 637–652.

Grote, G. (2012). Safety management in different high risk domains—All the same? *Safety Science, 50,* 1983–1992.

Grote, G. (2014). Promoting safety by increasing uncertainty—Implications for risk management. *Safety Science, 71,* 71–79.

Hale, A. (2000). *Conditions of occurrence of major and minor accidents*. Presented at the seminar "Le risque de défaillance et son contrôle par les individus et les organisations", Gif sur Yvette, France.

Heinrich, H. W. (1931). *Industrial accident prevention: A scientific approach*. New York: McGraw-Hill.

Helmreich, R. L., & Merritt, A. C. (1998). *Culture at work in aviation and medicine: National, organizational, and professional influences*. Brookfield, VT: Ashgate.

Helmreich, R. L., Klinect, J. R., & Wilhem, J. A. (2000). Assessing system safety and human factors with the Line Operations Safety Audit (LOSA). In *Proceedings of the Fifth Australian Aviation Psychology Symposium*. Aldershot UK: Ashgate.

Hofstede, G. (1980). *Culture's consequences: International differences in work-related values*. Beverly Hills, CA: Sage.

Hofstede, G. (1991). *Cultures and organisations: Software of the mind*. London: McGraw-Hill.

Hopkins, A. (1994). The limits of lost time injury frequency rates. *Positive performance indicators for OHS: Beyond lost time injuries*. Canberra: Australian Government Publishing Service, Commonwealth of Australia.

Hopkins, A. (2005). *Safety culture and risk: The organisational causes of disasters*. Sydney: CCH Australia Ltd.

Hovden, J., Abrechtsen, E., & Herrera, I. A. (2010). Is there a need for new theories, models and approaches to occupational accident prevention? *Safety Science, 48*(8), 950–956.

HSC (1993). ACSNI study group on human factors. 3rd report: Organising for safety. London: Health and Safety Commission, HMSO.

Hutchins, E., Holder, B., E., & Pérez, R. A. (2002). *Culture and flight deck operations*. University of California San Diego. Prepared for the Boeing Company under Sponsored Research Agreement 22-5003.

INSAG-1. (1986). *International Nuclear Safety Advisory Group's summary report on the post-accident review meeting on the chernobyl accident*. Vienna: International Atomic Energy Agency.

INSAG-4. (1991). *Safety culture. A report by the International Nuclear Safety Group*. Vienna: International Atomic Energy Agency.

INSAG-7. (1992). *The chernobyl accident: Updating of INSAG-1. A report by the International Nuclear Safety Group*. Vienna: International Atomic Energy Agency.

Journé, B. (2001, Juin). *Quelles stratégies pour gérer la sûreté? Le cas des centrales nucléaires françaises*. Paper presented at the XIe Conférence de l'Association Internationale de Management Stratégique. Faculté des Sciences de l'administration Université Laval, Québec.

Krause, T. (2012). PPT presentation by BST to ASSE Denver meeting 2012. Dislodging Safety Myths.

Kuhn, T. S. (1996). *The structure of scientific revolutions* (3rd ed.). Chicago, IL: University of Chicago Press.

Martin, D. (2013). PPT presentation by BST to ASSE fatality and severe loss prevention symposium (November 2013).

Maturana, H. R., & Varela, F. J. (1980). *"The cognitive process". Autopoiesis and cognition: The realization of the living*. Springer Science & Business Media. p. 13. ISBN 978-9-027-71016-1.

Merritt, A. C. (1993, October). *The influence of national and organizational culture on human performance*. Invited paper at an Australian Aviation Psychology Association Industry Human Factors Seminar, Sydney, Australia.

Merritt, A. C. (1996). *National culture and work attitudes in commercial aviation: A cross-cultural investigation*. Unpublished doctoral dissertation, The University of Texas at Austin.

Schein, E. (1990). Organizational culture. *American Psychologist, 45,* 109–119.

Schein, E. (1992). *Organizational culture and leadership* (2nd ed.). San Francisco, CA: Jossey-Bass.
Weener, E. F. & Russel, P. (1993). Boeing commercial airplane group, crew factors accidents: Regional perspectives. In *Communication to the 22nd IATA Technical Conference*.
Westrum, R. (2006). A typology of resilience situations. In E. Hollnagel, D. D. Woods, & N. Leveson (Eds.), *Resilience engineering: Concepts and precepts*. Aldershot, UK: Ashgate.
Wildawsky, A. (1988). *Searching for safety*. Somerset, NJ: Transaction Publishers.

Chapter 14
The Commodification of Safety Culture and How to Escape It

Hervé Laroche

Abstract Safety culture is a highly successful idea. Whatever your understanding of this idea, and whether you like it or not, you cannot ignore it. Safety culture has become a commodity (a product) that is promoted by various actors and enacted by various tools and practices. I first describe the 'safety culture system' that produces this commodification process. Then I discuss its upsides and downsides. Finally, I argue that, rather than debating on whether safety culture is a good idea or not, we should try to get the most of it by playing within the system that sustains the commodifying of safety culture. I suggest that safety culture should be taken as a vocabulary and as an asset. I also propose that rejuvenating the idea will come from introducing new actors into the system of safety culture.

Keywords (Safety culture) system · Commodification · Rejuvenation

1 Introduction

The concept of safety culture has generated a lot of enthusiasm but also a lot of criticism [for synthetic, yet opposite views, see Silbey (2009), Groupe de travail de l'Icsi «Culture de sécurité» (2017)]. Debates are still raging as to what the concept means exactly, whether it is useful or useless or even harmful. Founding texts are searched for exegesis, rather like sacred texts. What is to be considered as a founding text is also a debate, though. For instance, should it be the INSAG[1]/IAEA[2] or anthropological definitions of culture? These interpretive efforts are often supported by a historical perspective through which the genesis of the concept is retraced.

[1]International Nuclear Safety Group.
[2]International Atomic Energy Agency.

H. Laroche (✉)
ESCP Europe, Paris, France
e-mail: laroche@escpeurope.eu

C. Gilbert et al. (eds.), *Safety Cultures, Safety Models*,
SpringerBriefs in Safety Management,
https://doi.org/10.1007/978-3-319-95129-4_14

Definitions and genesis do matter, certainly, just not this much. Safety culture undoubtedly 'exists', first and foremost because many people do things in its name, whatever their understanding of the concept. In this chapter, I intend to describe what I call the 'safety culture system' and unpack its dynamics. The safety culture system is seen as a bundle of ideas, tools and actors that 'work' together and produce outcomes, such as safety culture assessments by auditors and consultants, safety culture training and change programs in companies, and safety culture literature by academics (this book itself is an outcome of the system). And of course, these outcomes themselves influence operational decisions and working behaviour in organizations. This is why understanding the system of safety culture matters: it translates into activities pertaining to safety and activities that have consequences for safety, which in turn will be understood in terms of safety culture and yield further activities to foster, repair, or amend safety culture.

This circular logic is known as 'performativity' in the social sciences (Gond, Cabantous, Harding, & Learmonth, 2016). The idea of safety culture is performative in the sense that it somehow enacts (produces) itself, or, rather, it enacts a world where something called 'safety culture' exists and is perpetuated. It is circular, yet it is not closed. I'll come back to this later. The key point is that it is, to some point, self-sustaining. Which is, as we will see, both a good and a bad thing. In any case, I will argue that this self-sustaining feature is what is to be taken into account by all those who seek to gain some reflexive capacity on safety culture. The concept of performativity has been key to understanding/explaining the development, hegemony and persistence of free market economics, for instance. I will build upon a conceptual framework put forward to account for the pervasiveness of rational decision-making theory and practices in today's organizations of all kinds (Cabantous & Gond, 2011).

2 The Safety Culture System

The safety culture system has three components:

1. the *concept or idea* of safety culture, and associated theories and models;
2. *tools and practices* that are used to develop and sustain safety in the name of safety culture;
3. *actors* who develop, promote and discuss the concepts and tools associated with safety culture.

The safety culture system thus encompasses the realm of ideas, the realm of activities, and the realm of entities (whether human beings or organizations). More important than distinguishing between the components is the understanding of the relationships between them. These relationships can be analysed as three different processes:

1. *conventionalizing* explains how the safety culture concept and associated ideas equip actors, and more specifically practitioners;
2. *engineering* refers to the process by which the safety culture concept is translated in tools and practices;
3. *commodifying* is the way by which tools and practices are diffused by and among actors.

The safety culture system can thus be pictured as a set of ongoing processes sustaining and reproducing the relationships between three components (Fig. 1). I will discuss these processes in more detail below.

2.1 Conventionalizing

Safety culture is a widely-held idea. It would be difficult to find anyone more or less involved in safety issues who would not be familiar with the idea of safety culture ("*safety what?*"). More importantly, many of these actors spontaneously think and talk about safety issues in terms of safety culture. Safety culture is thus a *convention*, in the sense that it provides a common ground for thinking and talking about safety issues. Many nuances, or even contradicting views of safety culture (what it is, what it does and how important it is) may coexist. Yet all these views revolve around a general idea and a set of associated constructs. When attacking it, the few adversaries of safety culture only contribute to the pervasiveness of the concept by stimulating counter argumentation. In short, they contribute to further conventionalizing the idea, that is, to making it a 'natural' way of understanding and acting upon safety issues. The all-encompassing, absorptive feature of the concept only helps with this conventionalization: safety culture is a flexible idea, its perimeter is rather vague, and other views of safety issues are allowed to survive either within or outside its perimeter (like probabilistic methods, for instance, or human factor approaches).

Fig. 1 The safety culture system

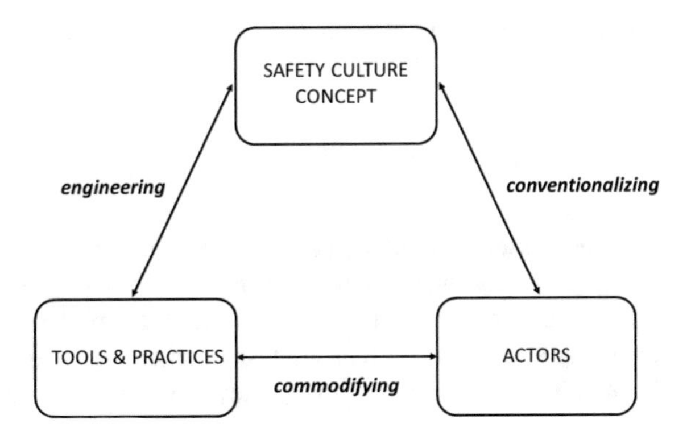

2.2 Engineering

A wide range of tools, techniques, approaches, and more or less standardized practices are available for diagnosing, measuring, characterizing, enhancing, maintaining, changing, etc., safety culture in organizations facing safety issues. Some originate from regulatory bodies and authorities, others are self-made, others are proposed by academics in books and articles, and many are marketed as products and services by consulting firms. This is engineering: building tools that operate the concept of safety culture. The performative power of such tools is easy to grasp: once a company has hired a consulting firm to assess its safety culture (whatever the company's managers think this means), it is bound to end up with a safety culture (now set out in a report), recommendations to improve it, an action plan to implement the recommendations, a set of tools to be implemented as a means of achieving the action plan, and another set of tools to monitor the implementation and measure its achievement. In this sense, the tools turn the company's safety culture into existence. This is not to say that a company's safety culture only amounts to these tools. Rather, the safety culture is made 'real' (understandable, actionable) by the tools that support it.

2.3 Commodifying

If safety culture has become a convention, then regulators, consultants, academics, and other practitioners have made it a commodity by 'selling' it to managers and organizations in search of safety. While engineering refers to turning ideas into tools, commodification refers to turning tools into products and services. Whether the market for these tools involves business transactions or not is of minor importance. Safety culture is indisputably a business. Yet it is also a commodity for academics who develop their career writing articles about it. It is a commodity for regulators who promote or impose it to industries. And Health & Safety managers also take it as a commodity when they advocate the roll out of enhancement programs to their top managers and ask for additional budgets.

3 A Spiral or a Circle

The conventionalizing, engineering and commodifying processes can be pictured as a spiral: once it has started, it grows and expands, each process reinforcing the others, each component drawing support from the others. Ideas are developed, discussed and refined. Tools are tested, amended, adapted. Actors appear and thrive by gaining legitimacy and/or money. Innovations are introduced by new actors relating to new ideas or new tools. New audiences are gained. Practices spread. Progress is made. This is the virtuous side of performativity.

It all comes at some costs, though. Conventionalized ideas are taken for granted and hinder innovative thinking. Engineering ideas into tools distorts the ideas, oversimplifies the issues, and turns them into technicalities. Actors become specialized technicians whose survival or prosperity depends on their skills at selling the tools they promote. Turf wars open between actors. Non-experts (e.g. managers) rely on experts without clear judgment. Tools confirm that tools are efficient. Exploration leaves place to exploitation. Commodification is everywhere. Top managers end up buying a new safety culture like they buy a new car. They give a lot of thought to buying the car, but not to the car itself (it's only a car, not a space ship). This is the vicious side of performativity.

As performativity develops, the spiral tends to turn into a circle, and circularity can become entrenchment. It should be clear that I am not seeking to apportion blame. Nobody really masters these processes and their outcomes. This is why understanding the processes matters: it is the key to regaining some degree of control over the processes.

4 Restarting the Spiral

Whether the spiral of safety culture has turned into a circle today is, in part, the topic of this book: the answer is unclear, yet the question is on the agenda. However, even if it were to be admitted that safety culture is on the verge of turning into a circle, the implications would still be open-ended. A drastic option could be, of course, to drop the concept, save a couple of tools, and let most actors sink or swim. Apart from the fact that no single actor, and especially not academics, has the power to trigger such a revolutionary turn, another, probably smarter, option might consist in restarting the virtuous spiral, rather than just bringing the vicious circle to a halt.

Restarting the virtuous spiral, or, less metaphorically, introducing innovative seeds into the performative dynamics of safety culture, can take several forms. This chapter does not pretend to draw up a list of them. More modestly, I will review the plausible sources of rejuvenation from three possible sources: ideas, tools and actors.

4.1 Ideas

Theoretical ideas about safety culture abound in the form of textbooks, handbooks and literature reviews. Even when reconsidering the basic terms, which is sometimes a good approach for stimulating new ideas, there is little potential. 'Safety' and 'culture' are well-established concepts. For instance, in organization studies, the concept of organizational culture is viewed as a mature idea (Giorgi, Lockwood, & Glynn, 2015). As Antonsen argues in this volume, there is no point in reinventing the wheel by developing new theories of safety culture. There is probably much more potential in using existing ones (or ideas deriving from the existing ones) that have been neglected or underexploited. Beyond this rather conservative position, I would like to make two suggestions to restart the spiral from the 'idea' side: safety culture as a vocabulary; and safety culture as an asset.

One interesting feature of the safety culture concept is that is has some plasticity. Rather than looking for more accuracy in the concept (its definition, its components, etc.), I would suggest making safety culture a flexible concept, to be adapted to the local circumstances of its use. Now that everybody is familiar with the notion of safety culture, let us take it as a vocabulary rather than as a theory. The vocabulary of safety culture enables actors to name and label things, thus removing ambiguities in complex sociotechnical systems. What is needed is a rich vocabulary, rich enough to remove ambiguities without simplifying too much. The vocabulary also enables actors to attribute causes to what happens or may happen. In the same vein, the safety culture lexicon should be varied and subtle enough so that attribution of causes does not lead to oversimplification. And finally, a shared vocabulary, obviously, fosters good communication and sense making. Safety culture vocabulary should enable communication within the organization (meaning, between operators, safety experts, managers, etc.), but also outside the organization (i.e. with regulators and the many sorts of stakeholders, including the public at large). The vocabulary of safety culture is to be constantly revised, enriched, and expanded.

Research about the culture concept in organization studies has produced at least one key finding: cultures are hard to change and thus cannot be used as a management tool in the short term. There is no reason why safety culture should be an exception. Yet, the commodification of safety culture, combined with the managerial obsession for change, obliterates this important feature. Safety culture is too often seen as a dependant variable that should be quickly adapted whenever a change occurs in the environment. An alternative view is to take safety culture as an asset. This is especially relevant, of course, for organizations with an 'advanced' safety culture. The painfully-acquired safety culture should be maintained, exploited, and developed, in order to stabilize the operational core. The implication being that it is the choices about products, technologies, structures, etc., that are adapted to safety culture. After all, this is what happens with financial capacities: they are treated as a constraint for strategic choices (or at least, they should be). Thinking of strategic choices in terms of what the safety culture can absorb does not condemn the organization to strategic inertia. Cultures cannot not easily be changed, yet they can absorb drastic changes when these changes are consistent with the key features of the culture (Ravasi & Shultz, 2006).

4.2 Tools and Actors

Reconsidering the existing tools and practices, amending them, and taking advantage of underexploited ideas to develop and test new tools and practices, would also appear to be a promising avenue. Though extant actors are, of course, a possible source of innovation, new tools and new practices are often best supported by new actors. This is why I treat tools and actors simultaneously.

'New actors' does not necessarily mean new consulting firms and new regulatory bodies. New actors may come from inside the organization. In the volume "Beyond Safety Training" (Bieder, Gilbert, Journé, & Laroche, 2017), several authors advocate for more empowerment of operators. Calling for the active participation of operators in the conception of technological systems, in the writing of rules, in the management of teams, and more generally in the monitoring of safety, is synonymous with introducing the operator as an actor in the system of safety culture. New tools and practices have to be designed to enable this actor to participate. Such a participation should in turn help for the redesign of existing tools and the development of new ones.

New actors may also come from outside the organization. The debate about the participation of external stakeholders or the 'public' is a complex issue that goes far beyond the scope of this chapter (Callon, Lascoumes, & Barthe, 2001). Yet safety culture is a powerful communication concept outside the organization. Dialogical practices could be developed under this umbrella and with the vocabulary of safety culture. The key question here is, whether or not organizations facing safety issues and regulators are open to these new actors.

5 Conclusion

Questioning the safety culture concept implies questioning the sociological system that produces and sustains safety culture as a set of ideas, associated tools and practices, and actors that promote and implement it. This system has turned the evasive concept of safety culture into a pervasive commodity. So far, this is a success, yet with some limits. The dynamics of this system are key for the development of safety policies in organizations. Our efforts, as academics, consultants, and practitioners, should be aimed at rejuvenating these dynamics, rather than refining the concept itself or the associated tools.

References

Bieder, C., Gilbert, G., Journé, B., & Laroche, H. (2017). *Beyond safety training: Embedding safety in professional skills*. Cham, Switzerland: Springer.

Cabantous, L., & Gond, J. P. (2011). Rational decision making as performative praxis: Explaining rationality's *eternel retour*. *Organization Science, 22*(3), 573–586.

Callon, M., Lascoumes, P., & Barthe, Y. (2001). *Agir dans un monde incertain. Essai sur la démocratie technique*. Paris: Seuil.

Giorgi, S., Lockwood, C., & Glynn, M. A. (2015). The many faces of culture: Making sense of 30 years of research on culture in organization studies. *The Academy of Management Annals, 9*(1), 1–54.

Gond, J. P., Cabantous, L., Harding, N., & Learmonth, M. (2016). What do we mean by performativity in organizational and management theory? The uses and abuses of performativity. *International Journal of Management Reviews, 18*(4), 440–463.

Groupe de travail de l'Icsi «Culture de sécurité» (2017). *La culture de sécurité: comprendre pour agir*. Numéro 2017–01 de la collection les Cahiers de la sécurité industrielle, Institut pour une culture de sécurité industrielle, Toulouse, France.

Ravasi, D., & Schultz, M. (2006). Responding to organizational identity threats: Exploring the role of organizational culture. *Academy of Management Journal, 49*(3), 433–458.

Silbey, S. S. (2009). Taming Prometheus: Talk about safety and culture. *Annual Review of Sociology, 35,* 341–369.

Chapter 15
A Synthesis

François Daniellou

Abstract This chapter aims to briefly summarize some of the key findings of the strategic analysis presented in this book. The main message is that a safety culture approach for an at-risk industry must be tailored according to what already exists in the company and to the aim that is pursued. Different historical backgrounds, different contexts, different constraints will require different ways and different paces for change and improvement. This short chapter suggests a number of prerequisites for a successful evolution in safety culture.

Keywords Safety culture · Safety model · Constraints · Taking stock
Strategy

The concept of "safety culture" has both an external and an internal use. When addressed to the outer world—be it the regulatory body, the media or the general public—its main value is to be as visible and brilliant as possible (a justification stake). When used internally, it is deemed to support the efforts made to improve safety. HSE departments are often caught between top management's expectations of homogeneous, company-wide, messages and approaches, and local management's needs for tailored and efficient support.

The first result of the strategic analysis presented is: "*If your main target is external, keep it simple, and use the models and formalisms that are the most popular for, or imposed by your interlocutors.*"

If your strategic goal is to mobilize internal stakeholders in the long term to improve safety, the question becomes: "*Under which conditions may the concept of safety culture help?*". Returning to the initial diagram in Chap. 2 (Fig. 1), it appears that the work of the strategic analysis leads to reconsider it in a rejuvenated way. Five dimensions are in tension with each other: a will or a need to act; an "already there", which is the influence of organizational culture on the ways of doing and the ways of thinking that affect safety; an offer (an academic and consultancy market)

F. Daniellou (✉)
FonCSI, Toulouse, France
e-mail: francois.daniellou@foncsi.icsi-eu.org

C. Gilbert et al. (eds.), *Safety Cultures, Safety Models*,
SpringerBriefs in Safety Management,
https://doi.org/10.1007/978-3-319-95129-4_15

of safety models; a set of stakeholders, for whom safety has to do with power; and the external environment of the group, the branch or the plant.

The dimension on which this book places the most emphasis is the need to properly understand and build on the "already there". The organization has other stakes to manage than safety, and its culture gives a certain weight to safety in the arbitrations made daily at all levels—which is reflected in the infrastructure: the technology and the organizational structure. But the present efforts towards safety are also borne by the collective practices of many professional groups, each of which has its own culture. A number of boundaries and interactions shape the landscape of safety culture.

The suggestion is to take this "already there" more as an asset than as a liability. It requires a careful understanding of the existing safety cultures, the identification of crucial trades and groups, and a deep understanding of "work as done" among them. It also entails the analysis of a possible "organizational silence" and of the reasons why information available in the field is not shared upwards.

Starting from this deep understanding of the "already there", conditions of success may be outlined. Suggestions for encouraging safety culture to evolve would be:

- Focusing on the prevention of SIF (severe injuries and fatalities) and major accidents, not on frequency rates.
- Involving all stakeholders at all steps of the process from the outset (the diagnosis and discussion of the existing situation).
- Examining what present safety cultures can absorb.
- Aiming to change their practices rather than their values—values will follow—while ensuring compatibility with the organization's overarching values and strategic orientations. Make visible efforts to reduce the everyday risks and optimize the working situations.
- Basing the change targets on the reality of activities, and on safety models that are relevant according to the endogenous and exogenous degree of uncertainty (one size does not fit all), and to the possible need for switching from one mode to another under certain circumstances.
- Regarding the change process, fostering meaningful dialogue and discussion, not only vertically along the managerial line but also, and maybe above all, horizontally at the boundaries between departments, groups, trades.
- Respecting the differences in perspectives, and endeavouring to reconcile the trades by providing a common vocabulary and representations favouring boundary crossing. Accepting a degree of ambiguity and friction.
- Holding together differentiation (according to the activities) and a common core of principles and indicators providing a clear framework.

A number of organizations are engaged in change processes aiming at improving their safety culture. A valuable contribution to knowledge about these processes would now be an in-depth analysis of the activities of the practitioners who bear the charge of carrying out the change. Behind the explicit announcements, which are the strategies, the trials and errors, the victories, defeats or rebounds? Understanding “work as done” should also be applied to change agents.

Afterword—A Number of Safety Models, Depending on Their Intended Use

After listening to the various international guests who attended the seminar and the discussions that took place,[1] I have come to the conclusion that there are at least four different schools of thought around the notion of a safety culture, that there is little interaction between them and that they are generally following different trajectories.

I do not seek to ascribe a value judgement to these different categories. None has definitively proven its primacy, and clients (the industrial sector) have long been used to picking and choosing from all of these visions, sometimes from more than one at a time, even if this generates a certain cacophony in terms of implementation.

A Shared Term, but Four Distinct Positions

These are the four schools of thought I believe I have identified:

The '**Pragmatics**', who take as their starting point the inappropriate behaviour observed in the field by the industrial client, often as an extension of a root cause analysis in accident and incident investigations. This is, I believe, the position of Dominic Cooper (Chap. 5). 'Pragmatics' offer consultancy services on the subject of Safety Culture. They do not burden themselves with theoretical debates on either definitions or concepts (indeed, to plagiarise what Binet wrote about intelligence: "*safety culture is what needs to be changed to reduce the risk*"), but they know how to listen to the people in the field and adopt a perspective that encompasses all industries:

[1]The two-day international workshop mentioned in the preface, organized by FonCSI in June 2016 and highlight of the project that led to this book (editors' note).

C. Gilbert et al. (eds.), *Safety Cultures, Safety Models*,
SpringerBriefs in Safety Management,
https://doi.org/10.1007/978-3-319-95129-4

> You have accidents, you have carried out in-depth analyses into the causes, you are aware of the defective HOF in your industrial system (diagnosis): problems stemming from management, behaviour or context.

The safety culture seeks to address these problems, proposing common sense actions on behaviour and organisations in order to reduce the risks of each company on a case-by-case basis, using an approach based on a belief model.

The '**Idealists**', who tend to focus on shared general values about what is and is not acceptable where safety is concerned. It seems to me David Marx (Chap. 7) belongs to this category. Once again, there is little credence given to academic debates about safety culture, nor to the detailed results of accident or incident root cause analyses. However, this group's proposals differ from those of the previous group. They tend to base their action on values to be adopted by all, offering little scope for opposition or falsification (as outlined by Popper): honesty, solidarity, transparency with regards to problem areas, equity of sanctions. The key to progress (in safety) lies in the capacity to (re)create communication around risk management within companies, as this communication is so often blocked by the fears of various parties. This approach shatters codes and uses a mixture of **legal and psychological strategies** that are easy and comfortable for the industry to understand, saying that it is normal to make mistakes, that many of these errors stem from the way competing pressures are handled (performance/safety; desire to do work well/desire to obey the rules). In doing so, the trick is to ensure strong spontaneous support from senior management (as the action impacts on the workers, rather than management, and the idea of punishing rule-breaking is maintained) as well as support from the workforce (because although we try to make people change, it is explained to them that their risk-making behaviour is the result of conflicts over which they have little control).

The '**Organisational Theoreticians**', such as Andrew Hopkins and Stian Antonsen, also put forward a proposal, but one that is different again to the two previous ones. This group starts with a theoretical pirouette around the subject: we agree with the idea of a safety culture but this is not the approach to take if one is to make progress in safety; rather the focus should be on the organisational culture, within which safety culture is just one of many facets. Consequently, we must:

1. **audit the organisational culture** rather than the safety culture, and
2. think more in terms of **competing priorities** (values, facets of the organisational culture) and work on these competing values if we want to change the aspect relating to safety. This aspect sometimes receives less attention compared to the other priorities, or is given enough attention at times and less at others, or is dealt with well in some places and less well in others (rapid variations in internal priorities in modern companies due to fact that they are spread out nationally and internationally).

One important point is that these authors are the only ones to introduce a socio-economic dimension of risk by considering that the organisational or company culture is also a very useful tool for forecasting the profitability and (economic) survival of the company, and safety is just one component of it.

The '**Sceptics**', who are critical of all the previous approaches. Mathilde Bourrier (Chap. 10), Gudela Grote (Chap. 9) and Jean Pariès (Chap. 13) would appear to adopt this position. They believe that the key question is elsewhere. The traditional approach to safety culture, which is designed 'retrospectively' once accidents have occurred in order to optimise the reduction of that particular risk in the future, has little chance of really reducing the risks of the future. Because these future risks stem more from unknown situations where procedures do not work or where operators are facing total uncertainty… (theories related to resilience) than from configurations that have already been seen in the past.

Here are some additional remarks.

It would have been difficult to construct this categorisation without direct contact with the protagonists, and without seeing how they actually interact with each other. Written documentation does not give access to this social 'intimacy' with people.

I was surprised by how secluded each expert seems to be, which reinforces my earlier analysis, particularly when observing that the experts in the various categories listed above have little to no contact with each other. And unsurprisingly, those who do know and see each other tend to share the same perspective of the four mentioned earlier. However, none had co-published with another.

The Roots and Variations of Culture

Debates between the participants provided interesting elements about what constitutes culture, and about the tolerance or otherwise of the dispersal of values relating to company culture within head office and in its national and international branches.

The foundation of company culture:

1. is largely dependent on its technical DNA and its national origins (to use a medical metaphor, we can say that these two factors are the 'genetic influence' part of the culture), but
2. it is also greatly (maybe more) dependent on the quality of the dialogue established within the company by senior management (starting from the highest level, Executive Committee as a model example) and the perception of trust present in all elements of the company (to continue with the medical metaphor, this is the 'environmental influence' of the culture).

It is normal for there to be a dilution of the intensity of the company's values in its national and international subsidiaries and one should not necessarily seek to strictly impose the same culture. However, there should be enough imposition through common regulatory constraints (common protocols, organisation and regulations) to limit this dilution.

The Question of Safety Models

Two key points emerge from the discussions. Firstly, a criticism of the excessive number of existing models. This means that we need far fewer but truly contrasted models, whereas models are being multiplied at will each time someone wants to make their voice heard in the community; in passing it is also a criticism of the emergence of resilience models, given that the very same ideas exist in HRO models.

The substantial contribution made by the 'sceptics' should also be highlighted. They quite rightly focus the initial debate on the safety model rather than on the content of the safety culture, with a very strong emphasis placed on the need for resilience and on the scientific inaccuracies of the traditional safety models based on the causal continuity of incidents/accidents (since Bird, there have been many similar models), and on the idea that reducing incidents also reduces the most serious accidents.

To Conclude

The idea that everyone is familiar with the expression 'safety culture' and that it is part of daily language (in the same way as expressions such as 'workload', 'awareness of the situation' and some others) is indeed shared by all the participants and means it is a precious tool for creating dialogue and for getting a foothold within the company (acting as a Trojan Horse).

The task that remains is to manage the interaction once inside the company, and the four points of view outlined previously give four very different approaches for doing this.

René Amalberti
FonCSI, Toulouse, France

Printed in the United States
By Bookmasters